Becoming Creole

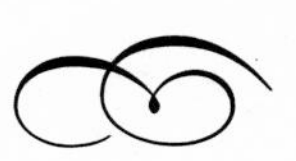

Critical Caribbean Studies

Series Editors: Yolanda Martínez-San Miguel, Carter Mathes, and Kathleen López

Focused particularly in the twentieth and twenty-first centuries, although attentive to the context of earlier eras, this series encourages interdisciplinary approaches and methods and is open to scholarship in a variety of areas, including anthropology, cultural studies, diaspora and transnational studies, environmental studies, gender and sexuality studies, history, and sociology. The series pays particular attention to the four main research clusters of Critical Caribbean Studies at Rutgers University, where the co-editors serve as members of the executive board: Caribbean Critical Studies Theory and the Disciplines; Archipelagic Studies and Creolization; Caribbean Aesthetics, Poetics, and Politics; and Caribbean Colonialities.

Giselle Anatol, *The Things That Fly in the Night: Female Vampires in Literature of the Circum-Caribbean and African Diaspora*

Frances R. Botkin, *Thieving Three-Fingered Jack: Transatlantic Tales of a Jamaican Outlaw, 1780–2015*

Melissa A. Johnson, *Becoming Creole: Nature and Race in Belize*

Alaí Reyes-Santos, *Our Caribbean Kin: Race and Nation in the Neoliberal Antilles*

Milagros Ricourt, *The Dominican Racial Imaginary: Surveying the Landscape of Race and Nation in Hispaniola*

Katherine A. Zien, *Sovereign Acts: Performing Race, Space, and Belonging in Panama and the Canal Zone*

Becoming Creole

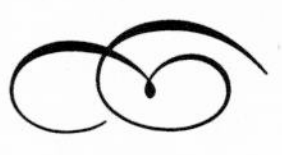

Nature and Race in Belize

Melissa A. Johnson

RUTGERS UNIVERSITY PRESS
NEW BRUNSWICK, CAMDEN, AND NEWARK,
NEW JERSEY, AND LONDON

Library of Congress Cataloging-in-Publication Data

Names: Johnson, Melissa A., 1962– author.
Title: Becoming Creole : nature and race in Belize / Melissa A. Johnson.
Description: New Brunswick, N.J.: Rutgers University Press, [2018] |
Series: Critical Caribbean studies | Includes bibliographical references and index.
Identifiers: LCCN 2017060242 | ISBN 9780813596990 (cloth : alk. paper) |
ISBN 9780813596983 (pbk. : alk. paper) | ISBN 9780813597003 (epub) |
ISBN 9780813597027 (web pdf)
Subjects: LCSH: Ethnology—Belize. | Racially mixed people—Belize.
Classification: LCC F1457.A1 J64 2018 | DDC 305.8/059602107282—dc23
LC record available at https://lccn.loc.gov/2017060242

A British Cataloging-in-Publication record for this book is available from the British Library.

♾ The paper used in this publication meets the requirements of the American National Standard for Information Sciences—Permanence of Paper for Printed Library Materials, ANSI Z39.48-1992.

www.rutgersuniversitypress.org

Manufactured in the United States of America

For my father-in-law, Alfred Elijah Bonner who loved history,
my sons Elrick Edward Bonner, Jr. and Adrian David Bonner and
my life-partner and cocreator of this book, Elrick Edward Bonner Sr.

Contents

Acknowledgments

I am indebted to so many who have helped me one way or another in bringing this book into being. First and foremost, I am indebted to the rural Creole Belizeans who warmly welcomed me into their communities, and who have since become my kin through marriage. The people who live in (or call home) Crooked Tree, Lemonal, Bermudian Landing, and the other villages of the Lower Belize River Valley made this book possible. In particular, my in-laws, the family of Alfred (Papa) and Lillian Bonner of Lemonal and their children (Elvis, Marlin, Christine, Elaine, Lessie, Elrick Sr., Claudett, Sharon, Shawn, Carolyn, Carlota, and Alfred Jr., their spouses and children), Orlando Banner and John (who graces the cover of the book) and Audrey Anthony, Ephraim (Mr. Big) and Olive and especially Bree Banner and Stephie Anthony I particularly want to thank Bree for reading one of the final drafts of the manuscript in full and providing much needed encouragement to me! And my adopted family in Crooked Tree, Rudolph and Gloria Crawford and their children, especially Dacia, Glenn, Dorla, and Rema, the family of Clifford and Gloria Tillett and their children, especially Sam, Dalia, Judy, Elvis, Bruce, and Lisa, and the Wades, especially Josephine, and also Miss Hazel, Wayne, Albert (and Pearl). Bernardine (Bo) Tillett, and really, *everyone* in Crooked Tree and Lemonal. Also, Annabelle and Eckert Guy generously housed me in Belmopan many years ago so that I could conduct archival research. My Belizean and Caribbean network here in Central Texas has also been important: Isola, Yolanda, Lily, Rose, Maurice, Al, Mr. Marcos, Judy, Jai, Denise, Swaggy, Ardette, Saraba, and others whose company (and food!) makes life here fun.

This project has been under way in various forms for nearly twenty years, and many different people have played important roles supporting me along the way. First, I owe my entire adult life, in many ways, to Richard Wilk, who introduced me to Belize in 1990, inviting me to work as his research assistant for a few months. I jumped at the chance, without even really having a good sense of where Belize was, and then decided that it was a perfect place to do doctoral research in environmental anthropology. Wilk also fostered in me ways of thinking that I now consider indispensable: a desire to question everything and assume nothing. My other professors from my graduate school days were also each helpful in important ways: Roy Rappaport, Richard Ford, Richard Tucker (who instilled in me a love for environmental history), Max Owusu (who made me think anew about whiteness), and Conrad Kottak all helped shape how I think. Funding for early research was provided by the Social Science Research Council, the Inter-American Foundation, and Fulbright-IIE program, and additional funding over the years has been generously provided by Southwestern University. I thank the library staff at Southwestern, and especially the interlibrary loan administrators for always swiftly providing any article or book that I needed. I thank Dean Alisa Gaunder for her steady support. An invitation to present some of this material at a Sawyer Seminar on Race, Place and Nature in the Americas at Rutgers University in 2013 was also very helpful. I thank Mia Bay and Ann Fabian for that invitation.

My network of fellow graduates of the University of Michigan Anthropology Program have been so important as supporters and critical thinkers over the years. Bridget Hayden, Bilinda Straight, Anne Waters, and Crystal Fortwangler have all been important interlocutors at different points along the way. And Coralynn Davis, John Stiles, and Deb Jackson have also been critical supporters over the years. Gina Ulysse and Michael Hathaway have been particularly vital to how this book has developed over the years and I am deeply grateful to each of them.

Fellow Belizeanists and environmental anthropology scholars: Laurie Medina, Jeremy Enriquez, Becky Zarger, Juliet Erazo, and Kareem Usher have also inspired and supported me along the way. And I remain forever grateful to Joel Wainwright for helping me settle on a title! Cristina Alcalde, a dear friend, and Timo Kaartinen both organized sessions at American Anthropological Association annual meetings that helped me develop my ideas.

My colleagues at Southwestern, Eric Selbin, Alison Kafer, Brenda Sendejo, Maria Lowe, Reggie Byron, Sandi Nenga, Omar Rivera, Ed Kain, Laura Hobgood, Romi Burks, Anwar Sounny-Slitine, Josh Long, Emily Northrop, Thom McClendon, and Jenn Esperanza (now of Beloit), have all been invaluable. Eric and Alison were important readers of the book, Brenda, Maria, Reggie, Ed, and Sandi have been helpful readers of early drafts of some of these ideas.

Rob and Jane Mackler, who share a deep love for Belize with me and who I first met when they were Peace Corps Volunteers in the early 1990s, have been staunch supporters over the years. I also thank Rutgers University Press editor Kimberly Guinta and her excitement about the book, and the expert work done by the team of people who have helped bring it into fruition (including Gina

Sbrili and Angela Pilliouras). The initial anonymous reviewers provided invaluable suggestions that vastly improved the book.

The students in my courses Global Environmental Justice; Race, Class and Gender in the Caribbean; Introduction to Cultural Anthropology; Theory in Anthropology; Capstone and Race, Nature and the Anthropocene have been absolutely critical to the development of this book. The stories that I shared in classes and my vision for the argument I wanted to make in this book developed in these places with them: the liberal arts college teacher-scholar model was at work here. My aim has been to write a book that would be accessible and interesting to rural Belizeans, introductory anthropology and environmental studies students, and academics interested in these ideas. To the extent that I have been successful, I owe my students.

Simone Yoxall and Caitlin Schneider, students at Southwestern, worked tirelessly to develop beautiful maps that also depict the area accurately and clearly.

My dear friend Julie Rocha, provided her unwavering faith in my ability to finish this project, celebrated milestones with me along the way, and was always there to listen to me. My neighbors Lonnie, Melissa, and Fynlee Syma and Conrad Frosch and family showed me something close to livity in my Central Texas backyard. My sister Carolyn Walton, brother Arthur Johnson, and their families, and especially Brighid Doherty have been sources of steady encouragement and engagement on these ideas.

My more-than-human companions over the years, black-and-tan coonhound (born in Belize), Shadow and then Sparky and Redbones Rosie, Luna and Rex, and Labrador Storm who all have walked across the rainbow bridge, and my current walking companions black-and-tan Webber and black Labrador Pepper encourage me to take walks along the San Gabriel River, entangling me with themselves and a host of more than human beings here in Central Texas, and providing space and time for my head to clear. And my cats, Kali and Peanut, both born in Belize who have since crossed the rainbow bridge and my current felid companions Mercedes and Sebastian who lie on top of papers, bite pencils and pens, demand attention, and thereby help keep me grounded.

Finally, my very biggest debt and acknowledgment is to Elrick Edward Bonner Sr., my life partner, husband, father of my children, touchstone and the soul who creates a world of laughter and fun for me. Elrick, known to more people as "Seedy," has talked with me about everything I have included in this book, and his insights shape what I see and know. I hope that our sons, Eldrick Jr. and Adrian, who took part in so many of the stories shared here, enjoy reading this book and learning more about their family roots in the Caribbean.

Becoming Creole

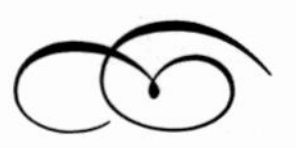

CHAPTER 1

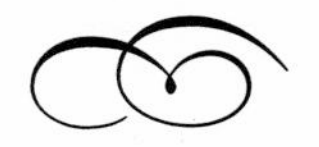

Introduction

BECOMING CREOLE

Irish Creek, a narrow and twisty waterway that winds through dense subtropical forest and empties into New River Lagoon, is a legendary hunting ground for rural Creole Belizeans. Distant from rural settlements, it is rumored to have the biggest halligatas (crocodiles), and densest concentrations of many animals (see Map 1.1 and Map 1.2). It is part of an area in Belize that has long been used by rural Creole Belizean men whose fathers, grandfathers, and ancestors before them hunted and fished when they were not working in the mahogany camps that were found here. During the dry season in March 2012, my husband, Elrick, was on his annual visit home to Lemonal, Belize, where he was born and raised. He joined a large group of his cousins and friends, all Afro-Caribbean men, most of whom lived in the rural village of Lemonal at the time, on an overnight hunting and fishing trip to Irish Creek. They were well furnished for this trip: many had camouflage clothing, there were tents, a boat and motor, nets, machetes; some of this Elrick had brought with him from States. They had fun traveling from the village of Lemonal to the bush of New River Lagoon where they set up camp and cooked the fish they caught on a small fire hearth they made. They were lucky this time and had found a number of hicatee, a type of turtle that is a favored dry season treat. They were replenishing their rural Creole masculinity—practicing skills and using knowledge that allowed them to thrive in these challenging subtropical wildlands. After camping on the edge of the large New River Lagoon, which is also the perimeter of the Rio Bravo Conservation and Management Area (RBCMA), one of Belize's largest protected areas, they set out on a morning hunt. As they started walking, they suddenly came upon a Belizean tour guide and warden for the nature reserve and his two white bird-watching middle-aged female ecotourist companions. These women had started their trip across the lagoon from the RBCMA headquarters, which advertises this place as wild and biodiverse. Humans are largely absent from the protected area's

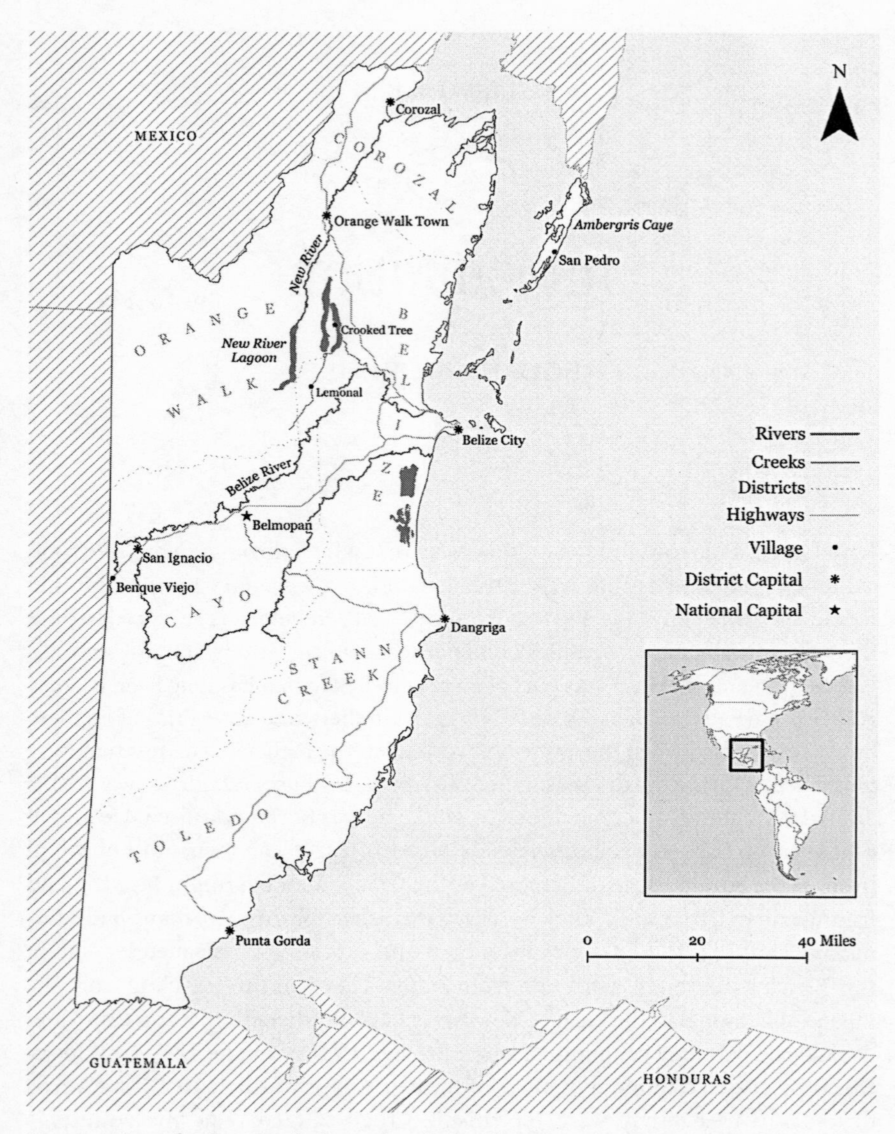

Map 1.1 Map of Belize. Made for author by Caitlin Schneider and Simone Yoxall, in Anwar Sounny-Slitine's Geographic Information Systems Laboratory at Southwestern University.

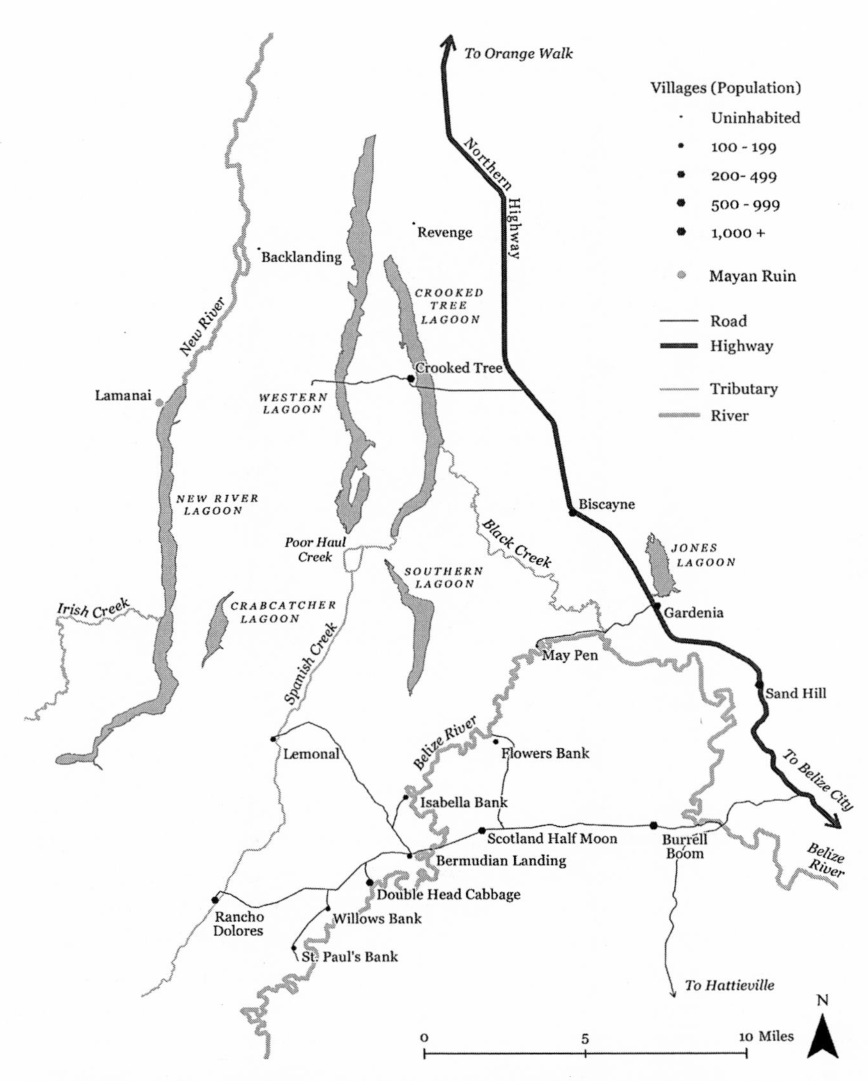

Map 1.2 Map of Research Area. Made for author by Caitlin Schneider and Simone Yoxall, in Anwar Sounny-Slitine's Geographic Information Systems Laboratory at Southwestern University.

promotional materials. When they are mentioned, the reference is mostly to the Mayan ruins found in the area. The ecotourists had not been expecting to encounter black men. During the many times Elrick recounted this story, whether over the phone to a cousin in Los Angeles, or around our kitchen table to visiting Belizeans in Texas, he chuckled and described what the two women had seen: "A big group of black men in camouflage with guns come outa noweh," he said, "and you could see their face change" (a Kriol way of saying they had a shocked or frightened look on their face) at the sight, and then he laughed, "I nearly shot a plug ovah dey head into the watah, but I didn't want to frighten the white ladies that bad" (I nearly shot a bullet over their head into the water).[1,2] The warden and his tourists continued on after the encounter, and the hunters continued on their day as well, catching two gibnuts, a large delicious rodent, on the trip as well. One year later, when I joined Elrick on a summer trip to Belize, we stayed in Crooked Tree and were visiting with an older Creole couple who have strong connections to both conservation and tourism. Elrick told the story again to the couple and a group of other people gathered around, including our then teenaged Belizean American sons. When he recounted all that he had hunted and caught during the trip, the older woman in the couple praised him to all who were there, saying, "He di do good, that da fi wi way of making a living from when wi pa mi young" (He is doing well, that is our way of making a living from when our fathers were young).

The encounter Elrick had, and its retellings, illustrate many of the themes I develop in this book. The historical connections between contemporary rural Belizeans and their woodcutting ancestors in this wild and watery place, the joy, camaraderie, and fulfillment that rural Belizean men feel when they are hunting and fishing together, and the challenges and dangers of being out in "the bush" are all essential to becoming Creole. At the same time, the possibility of blackened bodies encountering whitened bodies through entanglements of wildlife conservation and nature tourism, and the presence of conservation and tourism themselves are also now a critical part of becoming Creole. The telling of the story in Texas, to cousins in Los Angeles, and then again in Belize, on a visit to a different village, Crooked Tree, are all also part of becoming Creole for rural Belizeans today. The story also reveals my long and close connection to these communities and the intimate and auto-ethnographic approach I employ.

Becoming Creole brings into focus connections between the human, the more than human, and processes of racialization in Belize, a place that has articulated with global political economy since the sixteenth century. Belizean Creole people come into being through relationship with the more than human, or the material realities—the plants, animals, water, soils, and other entities—with which they live. These socionatural becomings are themselves entangled with processes of racialization in the context of transnational capitalist economic logics. Yet these logics are not all that shapes everyday life for rural Creole Belizeans. From the very beginning, Afro-Caribbean peoples in this part of the world have also generated and engaged in noncapitalist ecological and economic forms and ways of being in the world that might offer openings for imagining new futures in the Anthropocene.

Belizean Creole communities emerged through the encounter of enslaved "black" Africans and slave-owning "white" English woodcutters and an assortment of Englishmen too poor to own slaves and free "colored" and "black" men and women. These people all lived together in the swampy littoral of the Caribbean coast of Central America when this place was settled. Belize was simultaneously on the margins and integrated into global economies of timber extraction in the eighteenth and nineteenth centuries. In recent history, Belize has been a site for global processes of biodiversity conservation, ecotourism, and migration. Enslaved people and wage laborers of African descent developed their own economies and ecologies in this place at the same time that their labor fueled Belize's globally linked timber industry. Their descendants have continued to self-make as their homes have attracted the interest of biodiversity conservation and nature tourism and as rural Creole people have entered into circuits of transnational migration.

The river- and lagoon-crossed lowlands of north central Belize, with their tangled copses of tropical hardwoods, scrubby brush, and pine trees, and their dense populations of nonhuman mammals and reptiles, constitute a challenging and powerful landscape for humans to navigate. Belizeans' experiences of living among a host of powerful entities, of becoming who they are through their interactions with the more than human, and of creating economies not principally organized by capitalist logic, provide ethnographic evidence to substantiate analyses and critiques of coloniality and its discourses, like Sylvia Wynter's (2003) questioning of the Eurocentric foundations of the category of "the human," as well as to support the recent call for scholarship that is "attuned to life's emergence within a shifting assemblage of agentive beings" (Ogden, Hall, and Tanita 2013, 6). Belizean Creole socionatural assemblages show that human being is relational; people become who they are through their entanglement with each other and with agentive and powerful more-than-human entities.

The socionatural formations that entangle rural Creole people are co-constituted with processes of racialization. Blackness, brownness, and whiteness assemble with rural Creole people and the more than human in ways that sometimes support and sometimes challenge dominant white supremacist racial orders. Racial projects shaped the emergence of rural Creole people historically, and racial formations now assemble with biodiversity conservation, nature tourism, and transnational migration.

Most accounts of places like these emphasize how global forces influence and create local subjectivities. Yet I argue that Belizean Creoles have been creating alternative socionatures in the context of global processes since this place was first settled. Rural Creole people create a plurality of ecological and economic arrangements that articulate with, but are not overdetermined by, global forms of capitalism or bureaucracies of conservation. Here I provide an ethnographic grounding to the call for "re-mapping the familiar and opening ourselves to what can be learned from what already is happening in the world" (Gibson, Rose, and Fincher 2015, ii) in order to better negotiate life in the age of the Anthropocene.

More-than-Human Becomings

Theorizing human becoming as an entanglement with the more than human emerges out of several different scholarly developments in recent decades. Work in critical environmental theory, feminist, indigenous, and black theory and related strands in decolonial thought all contribute to the current florescence of scholarship. The idea of "nature" has been interrogated in Eurocentric social theory since the mid-twentieth century; theorists have pointed to the false dichotomy of human/culture versus nature, providing powerful analyses of how what has been imagined as "natural" is socially constructed (e.g., Braun 2002; Braun and Castree 2005; Castree 2013; Descola and Pálsson 1996; Haraway 2003; Merchant 1989). Bruno Latour's actor-network theory (Latour 2005), the related rhizomic thinking of Gilles Deleuze and Felix Guattari (1987), and Donna Haraway's cyborg thinking (1991) have suggested that those constructions must be understood as relational, emerging out of connection or assemblage (e.g., De Landa 2006; Escobar 2008).

A fruitful development in this line of thought has emphasized the point that nature is *not only* constructed, *not only a* representation, but that the world's vital materiality matters. Realities are constructed by physicality and embodiment as much as representations and ideas. Jane Bennett, in *Vibrant Matter*, works to bring human, nonhuman, and nonliving into analyses of being and becoming (Bennett 2010; see also Whatmore 2006). In an interesting analysis of the massive electric outage that affected a large area of the eastern United States in 2003, she identifies a myriad of ways that different nonhuman agents, "coal, sweat, electromagnetic fields, computer programs, electron streams, profit motives, heat" (Bennett 2010, 25), brought the outage into being. Or, in an example closer to the focus of this book, Laura Ogden beautifully describes how Florida's gladesmen, men who eked a living out of hunting alligators on the edges of the great Florida Everglades, became hunters, became who they were, through their relationship with water, alligators, mud, bullets, markets, buyers' desires for flawless skins, and more, in what she terms "the hunters' landscape" (Ogden

2011). In a later article, Ogden and her co-authors urge scholars to "understand the world as materially real, partially knowable, multicultural and multinatural, magical and emergent through the contingent relations of multiple beings and entities" (Ogden, Hall, and Tanita 2013, 6). Likewise, Sarah Whatmore asks scholars to see "the creative presence of non-humans in the fabric of social life and to register their part in our accounts of the world" (Whatmore 2002, 35). For all these theorists—who come from disciplines and persuasions as disparate as political philosophy, geography, and environmental anthropology—the world, and beingness, and becoming are always emergent and contingent and are best understood as entities coming together always "in the making" both materially and symbolically (Braun 2006).

Indigenous scholarship long precedes current mainstream academic thought on this front. Indigenous theorists' earlier work was rarely taken up by mainstream academics, and even today, many scholars do not acknowledge this groundbreaking work (Sundberg 2014; Todd 2016). Indigenous understandings of the world in particular recognize the agency of multiple nonhuman entities, material and nonmaterial: animals, plants, spirits, places, and other things all have the capacity to contribute to the creation and transformation of the world (e.g., Bawaka Country et al. 2015, 2016; Cruikshank 2005, Johnson et al. 2007; Lloyd et al. 2010; Rose 2004; TallBear 2015; Watson and Huntington 2008, 2014; Watts 2013).[3]

On Human Being

This relational understanding of human and more-than-human becomings shines a bright light on the problem of seeing the category of "the human" as a "biocultural given" (Ogden, Hall, and Tanita 2013, 7). The assumptions contained in our conception of the human facilitate misunderstandings and misrepresentations of so many people in the world, including rural Creole Belizeans. Sylvia Wynter's critique of the category of the human in Eurocentric thought, which she calls Man, provides powerful insight on this point (McKittrick 2006, 2015; Weheliye 2014; Wynter 1971, 2000, 2003; Wynter and McKittrick 2015). Wynter traces the development of the "ethnoclass (Western bourgeois) conception" of Man from its origin in the "Racism/Ethnicism complex" (Quijano 2000) that brought modernity into being from the end of the 1400s. This category recast prior Western European Judeo-Christian conceptions of the human as the "rational self" and the "law abiding political subject of the state" (Wynter 2003, 281). With the Enlightenment, this Western bourgeois ethnoclass of Man became "purely biocentric and homo oeconomicus." Biological science naturalized the racial order (among other dimensions of human being), and the economically self-interested rational subject was taken as the fundamental nature of human being.[4] Wynter's concern is that this very culturally specific, particularistic conception of the human, Man, has become "over-represented," as if *this and only this* is what it means to be human; rather than this category describing a small portion of the global population. Wynter contends that

> all of our present struggles with respect to race, class, gender, sexual orientation, ethnicity, struggles over the environment, global warming, severe climate change, the sharply unequal distribution of the earth's resources (20 percent of the world's peoples own 80 percent of its resources, consume two-thirds of its food, and are responsible for 75 percent of its ongoing pollution, with this leading to two billion of earth's peoples living relatively affluent lives while four billion still live on the edge of hunger and immiseration, to the dynamic of overconsumption on the part of the rich techno-industrial North paralleled by that of overpopulation on the part of the dispossessed poor, still partly agrarian worlds of the South)—these are all differing facets of the central ethnoclass Man vs Human struggle. (Wynter 2003, 260–261)

Wynter calls for a move beyond this overbearing conception of the human, she urges us to "speak instead of our *genres of being human*" (Wynter and McKittrick 2015, 31), the multiple forms of human being and becoming (always in a state of praxis) in the world (Wynter and McKittrick 2015, 34). These forms include human being, like rural Creole Belizean human being, that has been historically excluded from Man on the basis of European conceptions of race.

Theorizing Race and Racialization

Scholarship over the past few decades has well elaborated the ways in which race is socially constructed. This is to say that racial categories (e.g., "white," "black," "Asian") derive meaning socially and do not correspond to "natural" and "genetically distinct" groupings of humans (Gravlee 2009, Marks 2017).[5] The way in which racial categories acquire meaning and exert themselves in social worlds is understood best as historical and contextual. Michael Omi and Howard Winant's classic description of racial formation is still productive. A racial formation is "the process by which social, economic and political forces determine the content and importance of racial categories, and by which they are in turn shaped by racial meaning" (Omi and Winant 1986, 61). Many scholars have used this conception to investigate how race has emerged, operated, and been reproduced in a range of settings (e.g., Clarke and Thomas 2006; HoSang, LaBennett, and Pulido 2012; Joseph 2015; Simmons 2009; Thomas 2004; Warren 2001).[6]

In this book, I follow Arun Saldanha and others in conceptualizing race as an emergent and contingent process, or assemblage, that is simultaneously symbolic and material (Fanon 2008 [1952]; Saldanha 2006, 2007, 2012; Saldanha and Adams 2013; Wehileye 2014). While the meanings associated with race are socially generated, race is experienced and reproduced through physical and material embodiment and encounter (Fanon 2008 [1952]; Saldanha 2006, 2007, 2012, Saldanha and Adams 2013, Wehileye 2014) and race signs become encoded into the materiality of everyday practice as they assemble with differently positioned bodies (Nayak 2006, 418–419). Saldanha's use of the metaphor of viscosity to characterize racializing assemblages, with its potential for thickening and thinning, illuminates race's materiality and emergent character. What is commonly understood as "race" is a "configuration made viscous by a whole host of processes" (Saldanha 2006, 22). Particular ideas, phenotypes (specific physical embodiments like skin color and hair type), and things, such as hunger, wealth, beauty, rationality, and place assemble together repeatedly as "local and temporary thickenings of interacting bodies which then collectively become sticky, capable of capturing more bodies like them" (Saldhana 2006, 18; see also Nayak 2006, 418–419). As the same types of ideas, embodiments and things come together repeatedly, racial categories become more stable and more taken for granted as the way the world is; but when different ideas, embodiments and things come together, racial categories begin to break down. Thus, this conception of race as a viscosity recognizes the lived experience of how different embodiments move through the world, as well as the potential for mutability in racial assemblages.

Bringing a consideration of race into an analysis of human-environment relationships, or the more-than-human and human becomings of the world, builds on work that has analyzed race in relationship to place and nature. Scholars exploring these interconnections have described how places are both integrated into racial projects and racialized, and how people are racialized through their connections to place (Berry and Henderson 2002; Frazier, Margai, and Tettey-Fio 2003; Joseph 2015; Lipsitz 2011; McKittrick and Woods 2007; Razack 2002; Ricourt 2016). For example, urban ghettos in the United States are often racialized as black, whereas suburbs and rural poverty are racialized as white. Gina Ulysse offers an exemplary decoding of how different parts of the urban center of Kingston, Jamaica, are assembled with differently raced/colored and gendered bodies: light-skinned, slender upper-class "ladies" belong uptown, darker-skinned, heavier, working-class "women" belong downtown. This demarcation softened with the rise of dancehall queens and informal commercial importers that elevated the socioeconomic status of "women," and challenged the supremacy of "ladies" (Ulysse 1999). Rivke Jaffe shows how environmental ideas and practices intersect with the linking of race, color, class and place in both Kingston, Jamaica and on the small island of Curacao (Jaffe 2016). Rural places in Latin America and the Caribbean are also entangled in racial projects. For example, in Trinidad, East Indian and Afro-Caribbean Trinidadians each have specific associations with particular landscapes and lifeways. East Indians are more often imagined as rural agriculturalists, and Afro-Caribbean

Trinidadians are more often seen as urban dwellers, though these designations and associations are also always in flux. These place-based linkages have implications for the political and economic position and power of differently racialized people (Khan 1997).

Places designated as "natural" or "wilderness" also become enrolled in processes of racialization (Brahinsky, Sasser, and Minkoff-Zern 2014; Kosek 2006; Moore, Kosek, and Pandian 2003). Dominant discourse in the United States, for example, has dissociated African Americans from National Parks and other protected areas managed for recreational use (Finney 2014b; Glave 2010; Glave and Stoll 2006; Outka 2008). Parks and protected areas instead become co-constituted with whiteness, as do the nature and adventure tourism industries centered on them (Braun 2003; Fletcher 2014). Interestingly, at the same time, dominant racial discourse casts Southern and tropical lowlands and swamps across the United States and the Caribbean as sites of otherness; and that otherness is often shaded black (Allewaert 2013; Crichlow and Northover 2009a, 2009b; Wardi 2011).

Racialization of landscapes and peoples shape encounters over how places are used, including development and biodiversity conservation throughout the Global South. For example, in Latin America, state officials and others use racial discourses that cast indigenous and African-descended peoples land use as antithetical to development or as wasteful and environmentally damaging, in order to justify taking over their lands for a variety of purposes, including creating protected areas (Sundberg 2004, 2008a, 2008b; see also Checker 2008). Similarly, in the 1800s, places and peoples in Belize were racialized under colonial logic to ensure control of land and labor that favored the timber oligarchy that reigned there for the better part of two centuries (see also Johnson 2003). In recent times, racialization processes interpellate biodiversity conservation and tourism efforts.

The racialization of places where African-descended people live is informed by Eurocentric thinking that connected blackness and Africanness with beasts, and constructed black Africans as sub- and nonhuman. The association of animality with blackness was foundational to Europeans' invention of race in the sixteenth century, and was critical to separating whiteness from blackness in the early days of European imperialism and conquest (Jahoda 1999; Kim 2015). These associations persist in multiple cultural logics around the globe, and show up in how the "bush" is racialized in Belize and elsewhere.

In the analysis I present throughout the text, I pay attention to whiteness, brownness, and blackness and the racializing assemblages in which they are enmeshed. In European nations and white-settler states, whiteness includes a tendency to be invisible and unmarked (Bonilla-Silva 2013; Doane and Bonilla-Silva 2003; Faria and Mollett 2016; Twine, Gallagher, and Frankenberg 2008), even as it associates with political power and property, especially in the "raceless state" (Goldberg 2001). Whiteness's invisibility and normativity for white people associates with the "imagined modernity and rationality of Europe in the 19th Century and the US in the 20th Century" (Sundberg 2004, 54) and this assemblage enabled the development and reproduction of global white supremacy (Mills 1997, 2011). These associations informed the early development of social structures in Belize, as white colonial officers often held power, and this assemblage of whiteness continues to emerge as many international conservation organization representatives and tourists who visit Belize are white North Americans and Europeans, and as Afro-Caribbean Belizeans migrate to and work in the white-settler society of the United States.

Another relevant vein of scholarship interrogates colorism, a system of advantage based on skin shade and how distant or close one is to whiteness and blackness, both phenotypically and symbolically. How colorism works, the ways in which lighter skin offers advantage (or occasionally disadvantage) in different contexts, and how phenotype might be mediated by class, gender, and other material and symbolic entities, is a developing arena of inquiry (Glenn 2009; Norwood 2014). Analyses of how skin color and class intersect in the tourism industry in Cuba (Roland 2011) and how Dominicans are beginning to reclaim their African heritage after decades of denying and rejecting any connection to African blackness (Simmons 2009) help illustrate the complexities and nuances of how color operates in the twenty-first century, complexities and nuances that are equally present in Belize.

In *Becoming Creole*, I explore assemblages of race, color, and the more than human that first emerged in the context of the enslavement of Africans and the global capitalism that enslaved labor fueled. The systemic violence of this racial capitalism continues to generate racialized political economies in the twenty-first century (Pulido 2017), but racial capitalism does not fully encompass the world; there has always been the potential for liberatory assemblages of phenotypes (skin colors, hair types, facial features), things, and ideas to emerge (Nesbitt 2013; Slocum 2011). Carolyn Finney has issued a call for a new and different reading of race that acknowledges the role of political and economic power (often in the form of the state), but is still attentive and open to being human differently. In a rumination on race and blackness, she works to find "the creative and innovative ways that people expand and transform their realities" (Finney 2014a, 1279). Recent scholarship explores a variety of racial assemblages that offer liberatory possibility, or, in Sylvia Wynter's terms, encourage, the flourishing of multiple "genres of human being," from the critical political thought of black women in the United States (Cooper 2017) to emancipatory possibilities of black women's geographies (McKittrick 2006), Caribbean free villages (Slocum 2017), and the philosophy of Rastafarianism, including the increasingly used idea of "livity," which means a livelihood or way of living that includes being "free" (Roberts 2014, 2015). Wynter herself identified liberatory possibilities of blackness in the

Caribbean region she calls home. Carole Boyce Davies shows how Wynter imagined human being differently in the Caribbean Jonkonnu celebrations, festive masquerade street parades and drumming that occur with New Year's, or on other important days, and that date back to the earliest presence of African-descended peoples in the Caribbean. Wynter sees in Jonkunnu the possibilities produced in the most dire, exploitative, and dehumanizing conditions, and important to the thesis of this book, she locates these "alternative ways of human being" in place-based, grounded cultural production (Boyce Davies 2015, 220). Critically, these theorists do not analyze these formations of blackness as merely "resistance" to domination and oppression, but rather as alternate generations of being, formed not primarily in response to the prevailing order, but also on their own terms and logic. This possibility for liberatory difference animates my next theoretical interest.

Reading for Difference: Seeing Noncapitalist Economies and Ecologies

How the peoples and ecosystems of the tropics, and particularly the neotropics, have been affected by global capitalism has been well described and analyzed. The effects of resource extraction and industrial agricultural production on the peoples and natural environments living near these activities are often devastating (e.g., Li 2015). But it is not only these kinds of large scale industrial capitalist activities that have negative impacts Seemingly good efforts to protect biodiversity, or promote rural development through nature tourism, often reshape landscapes and remake cultures and identities in ways more harmful than beneficial (e.g., Brockington, Duffy, and Igoe 2008; Erazo 2013; West 2006). But in most of these instances, the peoples and communities experiencing upheaval have had some historical presence in the places where they live prior to being incorporated into regimes of capitalism.

African-descended communities in rural Belize have always already been entangled with global economic process, indeed these communities came into being as a result of European racial capitalism. And while the seemingly overdetermined nature of timber extraction, biodiversity conservation, nature tourism, and transnational migration might lead one to think that these communities are merely products of global process, this is not the case. Indeed, seeing these communities primarily in the terms of global process is limiting, and contributes to re-inscribing the "coloniality of being," as Wynter terms the rhetorical overrepresentation of Man.

In the analysis I pursue throughout the book, I attend to global logics of extractive capitalism and neoliberal regimes of conservation and tourism. But I also follow Wynter's exhortation to see that there is "always something else beside the dominant cultural logic going on, and that something else constituted another—but also transgressive—ground of understanding . . . not simply a sociodemographic location but the site both of a form of life and of possible critical intervention" (Wynter 2000, cited in McKittrick 2006, 123). Wynter pointed to the plot, which existed alongside the plantation in the sugar islands of the Caribbean (in Belize this was the "plantashe" that slaves and poor indentured people kept to make food for themselves). The folk culture of the plot was oriented toward use value, but was also a part of the plantation–plot dichotomy; this ambivalence constitutive of the black Caribbean subject (Wynter 1971; see also Mintz 1974).[7] In an essay elaborating Wynter's thesis about the plot, Elizabeth DeLoughrey notes that the noncapitalist orientation of enslaved Africans, who effectively transplanted and re-rooted themselves into Caribbean ecological assemblages after the horrors of the middle passage, and alongside the horrors of the plantation, is encapsulated in the word "nyam/n"—a pan-Caribbean African-origined word that means both "yam" and "to eat." Thus, the most foundational mode of self-care, nourishing oneself through eating, is named with a West African word and refers to a food with origins in West Africa that was also an essential food source for African-descended people in this new place (DeLoughrey 2011; see also Carney 2009; Carney and Rosomoff 2011).[8]

In *Ariel's Ecology*, Monique Allewaert has also explored this kind of possibility: the creation of alternative ways of being that emerge in the context of the black experience of enslavement and post-emancipation oppression, but that constitute their own creative agency on the edges of plantation slavery. She analyzes literature from the eighteenth century describing neotropical lowlands in the plantation zone and the people who live there, and finds African-descended peoples striving and thriving in interstitial and marginal spaces. They crafted relational ways of being and what Allewaert calls an "ecological personhood." This eighteenth-century phenomenon presages contemporary interest in ontology and the more than human: it is a relational ecology sitting alongside the modernist capitalist experiment of industrial agriculture and plantation slavery (Allewaert 2013).

The search for creativity, autonomy, and expansive human being in the apparently (but not) all-encompassing logic of capitalist slavery in the New World resonates powerfully with current theorizing on how to deal with the Anthropocene. Feminist political economists J. K. Gibson-Graham encourage us to find the radical heterogeneity of economic forms—capitalist and noncapitalist—that constitute our worlds, instead of assuming the primacy and omnipresence of capitalism (Gibson-Graham 2006a, 2006b, 2008). They describe in detail how the assumption that capitalism has fully "penetrated" all human economic forms everywhere misses so much economic activity *otherwise*, and is unwittingly caught in limited patriarchal metaphor and thought. Instead, they contend, people and things come into economic relation with one another in a host of alternative, noncapitalist ways that exist right there in broad daylight, but are simply not paid attention to. As Gibson-Graham turn their attention to the Anthropocene, they urge us to expand the search for noncapitalist economic forms to looking more broadly for alternative *ecological* forms. They call for an openness to seeing and feeling connection with each other, with the more than human, with "vital materiality" (Gibson-Graham 2011, 3). In the pages that follow, I aim to draw attention to these kinds of ways of living and being, these noncapitalist alternative economic and ecological forms that occur in rural Belize and the Belizean diaspora.

Worlding and Assembling

It is challenging to craft an analysis of more than human becomings in the context of, but not overdetermined by, capitalism that recognizes the materiality and social construction of race and that works toward "unsettling the coloniality of being." It entails recognizing the simultaneity of different "genres of human being" and more than human becomings. For example, it requires taking as equally real the very different becomings of a U.S.-based fisheries biologist and a rural Creole fisherman on a waterway in Central Belize. Mario Blaser and Juanita Sundberg provide some analytical tools to help with this. Blaser suggests that multiple overlapping and asymmetrically related ways of being are performed and enacted into being (Blaser 2010, 2014, 2016). These are "partially connected unfoldings of worlds" . . . "that interact, interfere and intermingle with one another" (Blaser 2014, 55). The fish-human worldings of the Creole fisherman and the fisheries biologist are simultaneous and overlapping, and are neither fully the same nor fully different. Furthermore, each account that people (scholars, or hunters, or anyone) create through their actions or their storytelling also contributes to bringing a particular world into being. And each account is also only ever partial. As the creator of this account, I am doing the same, contributing to bringing worlds into being. I must continually question the Euro-American worlding, the Euro-American ways of being that shape what I notice, what I think exists. Juanita Sundberg challenges us to proceed in this kind of scholarship by "walking and talking the world into being as pluriversal" (Sundberg 2014, 42).[9]

The concept of assemblage is useful in this endeavor, and also helps to keep in focus the materiality of these unfoldings. Tsing suggests that the concept of an *assemblage* asks us to think "about communal effects without assuming them" and helps us to see how "patterns of unintentional coordination develop" (Tsing 2015, 22). In an evocative study of Santeria, Aisha Beliso-De Jesús uses the concept of assemblages to encourage us to pay attention to the things people do, to materiality and physicality, and to nonfixity, or, as she puts it, to the "fluidity, intensity and dispersal and temporal-spatial impermanence of categorizations" (Beliso-De Jesús 2015, 8). In the pages that follow, I use the terms "assemblage," "entanglement," and "socionature" to capture the complexity, contingency, and materiality of relationships among multiple genres of human being and the more than human.

While I try to always keep processes of becoming human and multiple worlding clearly in focus, I also acknowledge here how difficult it is to write against the dominant frame of being and thought in which I have been raised. Trying to fully inhabit a way of being beyond assuming pre-existing fixed categories and bounded autonomous beings, and beyond the human–nature binary, is very difficult. In this book, I try to open my eyes to the connections people have with the more than human, to the noncapitalist, connected lifeways that exist in plain sight, to forms of liberatory blackness, and to multiple ways of human being. I adopt a stance of radical openness to the best of my ability (hooks 1990) and I look to my rural Belizean kin and friends as theorists and creators of knowledge who help me make sense of the assemblages and becomings in which they live (see Ulysse 2007, 6).[10]

Long-Term Intimate Ethnography and Generating Knowledge

Too often, scholarship on places like rural Belize is conducted and written by scholars whose subjectivities are nearly invisible, as if they are crafting universal truths through their writing (Ulysse 2007). Since the late twentieth century, indigenous, black, feminist, and other critics have been urging scholars to pay attention to how academic knowledge is produced, whose voices are legitimated and whose are not (Harrison 1991; Restrepo and Escobar 2005; Smith 1999; Weheliye 2014), but far too much scholarship leaves aside this important task.

The stories I share and analyses I put forward in this book were gathered during ethnographic research I carried out in Crooked Tree and Lemonal villages and their diasporas, between 1990 and 2017, including an intensive two-year period of field research between 1993 and 1995, as well as seven shorter periods of research in Belize between 1998 and 2013, and through social connection to rural Belize while living in Central Texas. In addition, I conducted extensive research in the national archives, and have reviewed a large number of newspapers, nongovernmental organization newsletters, and websites over the past twenty-five years. I am tightly tied to these people and this place, and not merely through repeated long-term ethnographic encounter. My connection is made much closer by my marriage to my rural Creole Belizean husband, Elrick, whose story opens this book. Our relationship began in 1993, when I was living in Crooked Tree and he in his home village of Lemonal, and has since included raising our two Belizean American young adult sons. I am thus tied not only by friendship, but by relations of kinship, of deep love and affection, as well as the complications that family ties generate, to many of the people and places I write about. Since my first two-year stay in rural Belize in the early 1990s, I have returned on average for four to eight weeks every three to four years—spending time in both Crooked Tree and Lemonal. My husband has returned every year since we moved to the United States in 1996, and since 2010, has been spending roughly three months each spring in Lemonal. When he is not there, he is in daily communication with people in Belize (and Belizeans in the United States) by phone and social media. I am also in frequent contact with my Belizean relatives and friends through social media. In addition, we spend free time with a group of Belizeans who live in Central Texas, and we exchange visits with Elrick's sisters who live outside Chicago. In many ways, I never leave the field: I am always deeply entangled with rural Belize and Belizeans, and these close and complicated connections inform what I relate in this book.

My methodological approach can thus be characterized as a form of "intimate" and auto-ethnography. The analytical themes and points the book makes emerge out of family members' and friends' stories and my own personal experiences (Walley 2013; Waterston 2013; Waterston and Rylko-Bauer 2006). My social location as wife, mother to Belizean American children, sister-in-law, daughter-in-law, Auntie, and cousin to numerous rural Belizeans intervenes in every encounter I have, always. At the same time that I am family to the primarily rural-born, non-elite Afro-Caribbean people I study, I am a highly educated white woman with a professional middle-class background from the East Coast of the United States. Being from the United States, white, a woman, and highly educated shapes and has shaped every relationship and social interaction I have. My attempts to make sense of what I have researched, seen, read, learned, and participated in are always entangled in these details of my gender, nationality, race, and class, and other dimensions of identity, in addition to my affine status—often in ways I may not even see, and almost always in conflicting and contradictory ways. I have been most often made aware of my whiteness and my U.S. nationality. Both of these identity markers, written on my body and through how I speak, more often than not have allowed me a whole array of privileges and facilitated my interactions with people; but at times I have also been reviled for that whiteness and foreignness. Because white often conflates with wealth and power, and authority, I was sometimes cast as an expert, or assumed to be connected to external conservation and development organizations. Likewise, the race, color, class, country of origin, gender, and age, among other identities, of the people I met and spent time with in Belize, shaped their encounters with me. The messy politics of those encounters mark everything that is in this book.

My outsider-insider status makes very clear the power dynamics that are at stake in the production of ethnographic knowledge and the codification of that knowledge into text form. Ethnographers like myself seek knowledge and understanding, and represent these findings most typically in written form. The people whose knowledge and understandings I sought are well aware of the value of what they offer. They choose to reveal and share what they know, or not (see Ulysse 2002, 2007). Rural Creole people are cautious about sharing information, and cultural and linguistic norms encourage this circumspection.[11] Indeed, my mode of often asking questions, despite everyone knowing that I was conducting social research and writing a book, was always a point of half-serious joking. During my first years of research and connection to rural Belize I was nicknamed by some "Channel 5," the name of a local news station, always with a laugh, but also always registering a discomfort. In those early years, other rural Belizeans were convinced I was an undercover U.S. Drug Enforcement Agency agent, or working for the CIA (Central Intelligence Agency). Over the past twenty-five years, this joking receded and my connection as family member has become the primary form of relationship I have with these communities. The rural Creole people whose lives I story in this book have afforded me a deep trust, one given more generously to me because of my family connections. My relatives are excited to know there will be a book about them, and trust that I will represent them in ways that feel comfortable and accurate to them. That they have this trust has made me take great care. I have tried my best to accurately and empathetically represent my family, my husband, my friends, and rural Creole culture generally in these pages.

Like Alisse Waterston, in her chronicle of her father's life as a Jewish émigré to Cuba, I have a "layered and complicated motivation" (Waterston 2013) in writing this book. It is not merely an ethnographic description of a group of people that I study, with questions in mind that I find intellectually interesting, but rather it is also my attempt to understand the world that I have chosen to be part of, through marriage, child-rearing, being a sister-in-law, daughter-in-law, Auntie. With this book, I aim to better understand myself and those whom I love. Part of my method has entailed sharing my analyses with Elrick, my sisters-in-law, cousins, and friends over the years, and sharing a draft of this manuscript with relatives who were interested in reading it. My goal in these pages is to "walk with" the people and places I describe (Sundberg 2014), and in that effort of walking with, I aim to adopt a posture of "radical openness" (hooks 1990). I want to be fully open to the potentialities of being affected by multiply embodied and constituted others (Roelvink 2015). Indeed, as I have tried to make sense of the differences I encounter in contexts of dedicated ethnographic research, or simply in navigating my marriage and making a home and raising children with my rural Belizean husband, I am often made aware of ways that I am not as open as I would wish, of ways in which I continue to have blind spots, even as I try to be open and learn. The longer that I am in this close relationship with these people and this place, the more I can see ways that I do not fully understand becoming rural Creole. The worldview and value system of my own racial, class, and geographic positionality are part of me much more than I want to admit.[12] As I move through this process of trying to understand, I see that I am in the midst of transformation (spiritual, mental, emotional) as a result of these engagements with "the visceral in the structural" (Ulysse, ginaathenaulysse.com), I too have been brought into being through encounter with the people I learn from in rural Belize as well as the material world of that place (Sendejo 2014).

Ethnography on Belize and Environmental Anthropology of the Afro-Caribbean

Becoming Creole joins a rich ethnographic literature on Belize, but a literature that has primarily focused on conservation, environment, health, and indigenous rights in Belize's Mayan communities (e.g., Baines 2016; Medina 2003, 2004, 2012, 2015; Wainwright 2008; Zarger and Stepp 2004), agricultural production in the banana and citrus industries (Medina 2004; Moberg 1997), and the Garifuna (Kerns 1983; Palacio 2006; Wells 2015). Richard Wilk's work is an exception. His monograph *Home Cooking* (2006) is based on ethnographic research throughout the country, with a detailed inclusion of Creole communities and culture (2006), and joins articles he has written on consumer culture (1989, 1990, 1995) and food (1999). Reflecting the long-historied dismissal of the New World African diaspora community as not having culture and being unworthy of study (Trouillot 1992), there is very little research on Belizean Creole culture: this book aims to fill that gap.

Similarly, environmental anthropology focusing on the Afro-Caribbean is relatively scarce, and tends to concentrate on coastal and marine issues, such as Amelia Moore's work on lionfish and the Anthropocene from the Bahamas (Moore 2012, 2016), David Griffith and Manuel Valdés Pizzini (2002) on Puerto Rican fishermen, and C. Guerron-Montero (2005) on marine protected areas in Panama. Exceptions include Crystal Fortwangler's research (2009, 2013) on invasive species, endemism, and conservation on St. John's, U.S. Virgin Islands, and earlier work by Sidney Mintz (1974, 1986, 2010) on sugar, yards, and agriculture. Rivke Jaffe's (2016) recent book on urban environments also takes environmental anthropology in the Caribbean onto new turf (see also Jaffe 2008). Analyses of Afro-Caribbean environments in geography include important pieces by David Lowenthal (1961) on Afro-Caribbean attachment to the land and Judith Carney's (2009) work on the African roots of Caribbean foodways.

Introducing this Place and these People

As map 1.2 shows, this book focuses on the northern central part of Belize. Two watersheds meet here: the Belize River (sometimes called Belize Old River) and the New River. The part of the area that interests me is called the Lower Belize River Valley, and I will sometimes refer to it as such. However, Crooked Tree Lagoon (and therefore Spanish Creek), also sometimes drains into the New River watershed, and for the people I describe below, New River Lagoon is as important, if not more important, than the Belize River. The landscape here is varied, from broad stretches of grassy savannah (that floods in the wet season) to high ridge—dense subtropical forest, to pine- and oak-studded pine ridge. Little of it is developed and large areas of land here support a diverse tropical flora and fauna. A number of protected areas have been declared here, including two that feature prominently in this book: Crooked Tree Wildlife Sanctuary and the Rio Bravo Conservation and Management Area, but the Belizean state is relatively weak, and the conservation sector is underfunded so conservation laws are not enforced as stringently as they could be.

This region of Belize is home to several thousand people living in a dozen or so villages with evocative place names: Crooked Tree, Lemonal (a shortening of Lemon Walk, "walk" meaning orchard or planted place), Rancho Dolores (ranch/farm of sadness), Double Head Cabbage, Bermudian Landing, Flowers Bank, and Scotland Half Moon, to name a few. These communities are located on the banks of waterways, either Crooked Tree Lagoon, Spanish Creek, Black Creek, or the Belize River, and were historically connected primarily by water. Many of these communities have only been connected by roads and bridges to the rest of Belize in the past thirty years. Similarly, the Belizean government has only run piped water and electricity to a number of these villages in recent years.

Many of the people who live in these communities can trace ancestry in this area back to the end of the eighteenth century. Crooked Tree and Lemonal, for example, were well established in 1830, but were likely inhabited by Belizean Creole people long before then. The people who live in these communities are all related through kinship ties, and those ties are closer between communities more closely connected by water (Rancho Dolores and Lemonal; Crooked Tree and Lemonal; Willows Bank and Bermudian Landing, etc.). Particular surnames are associated with different villages. For example, Crooked Tree surnames include Tillett, Crawford, Gillett, Wade, and Rhaburn; Lemonal is home to people with the surnames Bonner/Banner, Anthony, and Sutherland, among others; McCulloughs live in May Pen; and Flowers in Flowers Bank.

The villages are ethnically Creole. Belize has a number of ethnic groups: Creole, three groups of Maya (Kekchi, Mopan, Yucatec), Garifuna, Mestizo (this category includes Spanish-speaking Belizeans who have long ancestries in Belize, many of whom also sometimes identify as Yucatec Maya, as well as a growing number of migrants and refugees from Guatemala, El Salvador, and Honduras), people of East Indian/South Asian descent, and a relatively large population of Mennonites (German-ancestried members of an Anabaptist religious order). Belizean Creoles are descendants of a mix of Englishmen and enslaved and free people of African descent, though many Creole people also have Mayan or Mestizo ancestors. The Creole population has been politically and culturally dominant in Belize, though recent growth in the Mestizo population and emigration of Creole people to the United States may be weakening this dominance.

Belizean Creole people identify as Afro-Caribbean, and feel a strong sense of connection to Jamaicans, Trinidadians, Barbadians, and other Afro-Caribbean peoples. They share foodways (rice and beans), music (reggae, dancehall, soca, and calypso) and festivals (carnival, Jonkonnu, and more) with the broader Caribbean. Belize's national language is English, and many Belizean Creole people feel a close affinity with England and cherish the Queen. But the lingua franca throughout the country is Belizean Kriol, which is an Afro-Caribbean Creole—with origins equally in Europe and Africa. In the case of Belizean Kriol, the mix is between English and a number of West African languages. Kriol is spoken along a continuum of more English/less African to less English/more African.[13] Belizeans refer to speaking Standard Caribbean English as "speaking," otherwise people simply "talk." If Standard Caribbean occupies one end of the continuum, Belizeans call the other end "broad" or "raw" Kriol. Like other Caribbean Creoles, Belizean Kriol is replete with double meanings, and talking Kriol is often a self-conscious, performative, and competitive act (Abrahams 1983).

People here make a living in a wide variety of ways; many, if not most, of the people who live in these communities participate in a variety of activities to sustain themselves. Regular bus service allows people to commute to jobs in Belize City, Orange Walk, or Belmopan. A small-scale nature tourism industry is well established in this part of Belize, and the various tourism-related enterprises are most often owned and managed by people from these communities.

As elsewhere in the Caribbean, this region of Belize has experienced a great deal of emigration to the United States over the past several decades. In addition, people migrate within Belize: moving between villages, and from villages to cities, and vice versa. Every Belizean is a member of a usually large network of kin spread across Belize and the United States. Remittances sent by relatives in the United States back to these villages, and plans for return migration as well as the migration itself provide a steady influx of money into the villages from the United States.

A Preview of What Is to Come

In chapter 2, "Hewers of Wood: Histories of Nature, Race, and Becoming," I describe the historical roots of rural Creole presence in this place. I begin with the emergence of the socionatural in these swampy lands from the days of early Mayan presence to the seventeenth-century arrival of (mostly) Englishmen.

Early woodcutters were drawn by abundant logwood with its rich purple and blue dye, and kept here by the promise of wealth from the more labor-intensive mahogany industry, which, to turn a profit, required the labor of enslaved Africans and African-descended people. The enslaved Africans and both their free and enslaved descendants were violently deracinated from their places of birth and ancestral homes but re-rooted themselves through their relationships with the more than human in this place, alongside Englishmen, who for the most part chose to live here. In the following centuries, African and European together created a place-based culture. The socionatural assemblages that constituted this re-rooting shifted through time and were entangled in emerging logics of race, color, class, and place shaped by a broader racialized political economy. Belize's race, color, and class hierarchy emerged alongside the identity Belize Creole in the late nineteenth century. As much as African-descended people labored in the timber industry, or, in the mid-twentieth century, partook in conservation and tourism developments, they engaged in alternative ways of making a livelihood, from fishing to hunting, raising cattle, selling crocodile skins, serving as tour guides, and conducting a host of other activities that echo the occupational multiplicity and flexibility that characterizes Afro-Caribbean being. The linkages and assemblages described here leave traces that are taken up in the following chapters.

In chapter 3, "Bush: Racing the More than Human," I further develop the theme of racialization as I consider how Belizean Creole relationships with the natural world exist within global discourses of race and being "backwaad." Landscape aesthetics and ideas about the bush—or the natural world beyond a village's edges—reflect the articulations of global discourses of race with the relationship between the more than human and human in Creole communities. I closely examine the very frequently used words "bush" and "bushy" for the different ways they are used to describe people and places, and the different ways they index race. I use the rural Belizean yard, and its capacity to be bushy, as a central point of analysis. People work hard to keep certain nonhuman elements at bay in their yards, and they evaluate places, and the people connected to them, in terms of how bushy they are, or how clean they are, how well the more-than-human elements are kept out. The idea of bush as an evaluation is double-sided and ambivalent—used strategically depending on the context as a marker of superiority and inferiority, and always contributing to racializing assemblages.

Chapter 4, "Living in a Powerful World," explores how rural Belizeans experience this place, a place they share with agentive, sapient, and often dangerous more-than-human entities. From the power of water to the dangers and capacities of a range of nonhuman animals (fer-de-lance, jaguar, crocodile) and the more-than-material beings found throughout the region, the occupants of these places challenge the centrality of human beingness in the world. These worlds preclude conventionally understood human mastery and control. Nonhuman agency and presence is clear to rural Creole people in ways that are elided in dominant Eurocentric ways of being.

In chapter 5, "Entangling the More than Human: Becoming Creole," I examine how the relationship between humans and the more than human generates what it means to be human—in this case, to be rural Belizean Creole. I show how the more than human is central to creating the cultural meanings and social networks that make rural Creole culture and community; rural Creole Belizeans are both folded within and derive identities from intimate connections to the more than human. Rural Creole people come into being through practical engagement with the natural world: fishing, turtling, hunting, making plantashe, gathering fruits, raising cattle, cooking, preparing game meats, and so on. Networks of sharing and exchange create Afro-Caribbean community, but in rural Belize, it is the sharing of these "natural" goods that are most important. These exchanges of fish, deer, peccary, gibnut, fruits, to name a few items, are central to the noncapitalist economies that rural Creole people have been creating since the seventeenth century, and that in turn generate Creole culture and community. In addition, the more than human serves as the primary source for metaphor and symbolism in the linguistic play of the Creole language, in storytelling and in Belizean musical production.

I consider the impact of conservation projects and ecotourism on rural Creole socionature and racialization in chapter 6, "Wildlife Conservation, Nature Tourism, and Creole Becomings." Global momentum to protect nature coincided with the popularization of tourism in the mid-twentieth century and Belize was caught in these currents from the onset. Crooked Tree was one of the first places in Belize where these currents converged. This part of Belize, with its seasonally flooded lowland savannahs and their endless rivers and enormous flocks of every kind of waterfowl imaginable, including the six-foot-tall and rare jabiru stork, was an early attraction for white U.S.-origined ecologists and birders, and led to the creation of a wildlife sanctuary and ecotourism (especially bird tourism) hotspot by the early 1980s. Conservation and tourism thus became entangled with rural Creole Belizean socionature. Rural Belizeans took advantage of new ways to engage the more than human, and also defiantly continued to generate alternative ecologies. Race and the logic of white supremacy powerfully organize both biodiversity conservation in the Global South and tourism (typically from northern countries to southern countries); and rural Belizeans are highly attuned to the ways in which they and their communities appear cosmopolitan (or not). The encounters provided by ecotourism and conservation at different points in the past forty years offer moments for reflection on what it means to be human among the nonhuman, and how processes of racialization intervene in these relationships.

In chapter 7, "Transnational Becomings: From Deer Sausage to Tilapia," I analyze the circulation of beings from their places of origin to other places and the socionatures that emerge from these movements. I begin with a focus on the migration of Belizean Creoles to the United States and how the more than human is central to generating a sense of being Belizean for migrants living in the United States. Belizean Creoles re-create being rural Creole through bringing Belizean bush foods into the United States, and having gatherings and celebrations centering on those foods. Or, they engage in hunting and fishing in the United States, and cook and share this U.S. game and fish to generate rural Belizean connection. Gatherings are posted on and celebrated through social media. Migrants also return, to live or to visit—often annually, and through these returns transnational Belizeans replenish their rural Creole being. During their visits, they deeply engage with the more than human of rural Belize, and infuse resources and money into the community, shifting socionatural processes. Through all these activities, socionatural assemblages create community across the diaspora. Another movement across borders that has reshaped rural Creole Belize is that of tilapia. Introduced into Belize for aquaculture in the early 1990s, tilapia then escaped into local waterways during flooding. These escaped fish were initially rejected by rural Creole people but eventually were embraced as central features of rural Creole economies and ecologies.

Chapter 8, "Conclusion: Livity and (Human) Being," recapitulates my main arguments, while considering how thinking with rural Creole Belizeans helps us better engage contemporary issues such as the crisis of the Anthropocene, through rethinking what it means to be human. It highlights the way in which understanding how rural Belizeans self-make and world-make in ways not overly determined by the global political economic and racialized processes in which they are entangled might provide insight for how to live well in the world.

CHAPTER 2

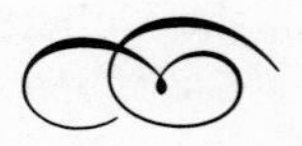

Hewers of Wood

HISTORIES OF NATURE, RACE, AND BECOMING

Cousin Ezekiel is sitting across from me on the porch of his small high house in the center of Crooked Tree Village on a sunny morning in the dry season in 1994.[1] He lives alone here now, at the tail end of his life. He is still a formidable figure, six feet tall, with piercing blue eyes, and a thick shock of white hair framing his "clear," or light brown, face. He looks out across the roadway, across his "bushy" yard remembering years past, telling me about how he used to leave the village to work at the mahogany camps of Belize Estate Company, about how his brother left Belize for Scotland when they were young, and never returned. He talks about hunting and fishing, about how high the bush was in the village, and the "tigahs" that were always on the outskirts. I am fascinated, sitting on the porch with my "sister," who accompanied me out of both kindness and interest. She had heard these stories a million times. Stories told by the "old heads" in rural Belize filled more evenings than I can recount, although they seemed more plentiful when the village generator was broken and the bright moonlight beckoned people out onto benches propped up against houses. Cousin Ezekiel was enjoying talking, laughing at old exploits, enjoying thinking about his youth here, and we loved hearing his stories.

The history of this place and of the ancestors of the people who live here today, ancestors who made livelihoods here for hundreds of years, is an everyday topic of interest. Social gatherings are peppered with remembrances of who came before, how people are related to one other: "so, fi he pa mi haffi call Mr. Jonny uncle?" (so, his father had to call Mr. Jonny uncle), and tales of times long gone. This chapter honors the stories I heard and lays out the historical record, scant as it may be, that further substantiates them.[2] The history of this part of Belize shows how people have been self-making with the more than human here, in the context of a racialized capitalist political economy, for centuries.[3]

The Maya(cene)

This place has never been in stasis, the more than human itself an agent of constant change. Heavy rains flood the area during parts of the year, intense hot sun dries it out at others. Grassy savannahs become flooded lagoons, crocodiles create furrows in mud, marching army ants cut wide paths across forests—this land of lagoons, creeks, savannahs, and dense copses of tropical trees is always in the process of becoming. Humans have been part of these processes for several thousand years. Archaeological evidence suggests Maya peoples lived in Belize from at least 2000 B.C., the civilization emerging, flourishing, and then again collapsing around 1500 A.D., with smaller settlements and cities occupying this part of the isthmus through the present (Campbell 2011).

The Maya built a number of large ceremonial and political centers, one of which was the Lamanai complex on the New River, just a few miles from Crooked Tree and Lemonal. There are also smaller sites throughout this area, including Chau Hiix (Maya for jaguarundi, or hallari, in Belizean Kriol) on the edge of Western Lagoon. The landscape of the northern half of Belize is replete with evidence of agricultural and residential sites of Maya peoples. Rural Creole people know numerous "Indian hills" and these are increasingly being identified and mapped by archaeologists (Beach et al. 2015; Dunning et al. 2002; Lucero et al. 2014; see also Kawa 2016 for a similar argument about the Amazon). The lagoons in this area show evidence of extensive dams and canals, indicating Maya hydrological manipulation for agricultural purposes (Harrison-Buck 2014; Pyburn 2003). These mounds typically contain dense stands of fruit trees and other useful plants (mammy apples, for example), that are otherwise not found in such concentration. While this area was clearly densely settled by Maya in pre-Colombian days, the number of Maya living here when Spanish first crossed these lands and founded cities in the Yucatan in the 1500s, and then when British buccaneers began using these lands, eventually creating small settlements in the 1600s, was likely smaller. These Maya, however, did not simply acquiesce to European presence. Archival materials reference periodic raids made by groups of Maya over the ensuing two hundred years. From the earliest British settlements, archival records reveal concern about the "Wild Indians," as they were named by colonial officials who feared their raids, and show how the British racialized Maya as an ultimate "other." Yet, there was also intermingling between Maya and the English and African descended people who lived here. Many Belizean Creole people today can identify a Maya person in their ancestry. and the foods people in rural Belize eat and plant-based medicines they use have traces of Maya influence.

Buccaneers along the Coast, 1500–1650

In the sixteenth and seventeenth centuries, European countries expanded their political and economic control globally as mercantile capitalism strengthened. The New World of the Americas offered dreams of El Dorado and heretofore unknown riches to the Europeans who explored them. These dreams lured many European men to the New World. The combination of the promise of great riches, the attenuated presence of state power, and squabbling between European states seeking control of the area fostered piracy throughout the Caribbean and coastal Central America. The European men (and most were men) who came to the Western Caribbean at this time were sailors and buccaneers, hustling for ways to get rich and choosing to live at the edges, where state power struggled to reach. Although the coast of Belize offered protection because of its shallow, reef-filled, and caye-dotted waters, the area was also difficult to navigate and was located too far from the main trade routes to be an ideal spot for pirates. Sailors who spent time along this coast in the 1600s likely relied less on piracy and more on gathering goods from the shallow seas, such as turtle shells and pearls to trade, and thereby developed a growing knowledge of and connection to these coastal environments (Bulmer-Thomas and Bulmer-Thomas 2012). When privateering became illegal in 1667, some buccaneers turned to a more settled and staid way of making a living: cutting wood to sell. For these men, choosing to live, or better put, squat, on the margins, in lands claimed by Spain and ignored by Britain, reflected their renegade propensities and their interest in being left alone

Logwood's Liveliness and the Early History of Belize, 1650–1790

The wood the buccaneers sought was the increasingly valuable logwood tree, from which a purple and red fixing dye was extracted for use in the rapidly growing textile industry in Britain (Bulmer-Thomas and Bulmer-Thomas 2012; Campbell 2011; Wilson 1936). Logwood grows abundantly in dense stands along the edges of the plentiful rivers and other waterways that stretch across the swampy lowlands of northern Belize. It is a relatively small tree, indeed it might be better called a shrub, and belongs to the bean family.

The process of cutting and transporting logwood is relatively simple and did not require much labor. Belize historically has had two dry seasons each year, one in March–April, and a shorter one in August, during which wide and shallow lagoons are transformed into dry savannahs of sweeping grass, and only the deepest parts of creeks retain water.[4] During dry seasons, logwood cutters set up camp close to shallow waterways where logwood grows, and cut the short and stocky tree, piling the heavy trunks into heaps and digging out a path to each heap. As the rains returned at the beginning of the wet season, the pathways

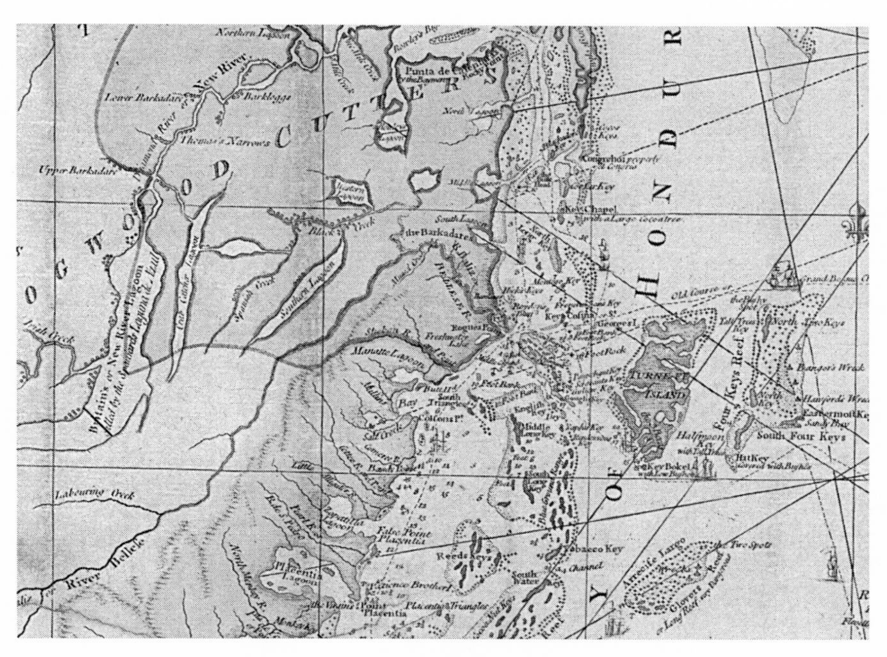

Figure 2.1 Close-up of 1775 map of Bay of Honduras, showing centrality of this area to Belize's history. *The Bay of Honduras*, by Tho' Jefferys, Geographer to His Majesty, London, 1775. Source: British National Archives, Kew.

flooded and the log piles could be loaded into canoes, and canoes floated down the now flooded path to a major waterway, and then to a "barcaderes" (landing place) on the Belize River, which is always deep and running.[5] In early accounts of logwood-cutter life the most commonly used landing place was likely the "Old Barcaderes" marked on a map drawn in 1775 (see figure 2.1), and located downriver from the junction of Belize River and Black Creek, the tributary that leads to the lagoons at Crooked Tree and Spanish Creek.

After floating down their piles of logwood "sticks," these cutters spent the wet season at the Barcaderes. From here, they traveled swiftly down to the Belize River mouth if they heard word that a ship had arrived, returning to the Barcaderes loaded with liquor and provisions. The woodcutters enjoyed themselves until the next dry season, when they again dispersed to their "works" and "banks" up the rivers and creeks and along lagoon edges.

Living in logwood-cutting camps and at the Barcaderes necessitated deep entanglement with the more than human in these places. Woodcutters negotiated the ebbing and flowing waters, torrential rain, scorching heat, mosquitoes and sandflies and hunted a wide variety of game, ranging from large wild birds (quam, currasow) to deer, turtle and a variety of game meat (Wilk 2005). They collected fruits that this place offered, from avocado to mammy apple, grew plantains and tubers like coco and cassava, and constructed homes, tools, and other items from the timber and palm trees they lived among (see Craig 1969; Joseph 1987, 1989; Wilson 1936). These early buccaneers incorporated themselves into the socionatural assemblages that would generate Belizean Creole culture.[6]

In addition, from the moment that Europeans first sailed to the Americas, they released livestock they had brought with them from the Old World wherever they landed. Cattle, horses, goats, sheep, pigs, and chickens were brought very early on to Belize, as were plantain, rice, okra, and a variety of other plants and animals native to the Old World. While Europeans intentionally introduced a host of new species into the Belize, others hitchhiked and found themselves on a new continent. These include the notorious *Aedes aegypti* mosquito and its capacity to carry yellow and dengue fevers, and the deadly malaria plasmodium that traveled in European and African human bodies. Whether domesticated and husbanded carefully by their human transporters or freely roaming and dispersed, all these forms of life joined the weedy species of humans as invaders into the New World (Crosby 2003 [1973]; McNeill 2010; Subramaniam 2014), invaders who would make themselves a home here, joining the always-emergent socionatural assemblages that I trace in this book.

Hewers of Wood: Creating Alternative Economies and Polities on the Margins

In *The Many-Headed Hydra*, Peter Linebaugh and Marcus Rediker argue that throughout the Atlantic world between 1500 and 1750, while capitalism was developing as an economic system, alternate and oppositional forms of economic and political organization were being created on the margins (Linebaugh and Rediker 2013 [2001]). These alternative economies emerged among pirates, on some sailing ships and in other places that evaded or did not attract the full interest of European states. The early days of logwood cutting and settlement in Belize are one such example.

The buccaneers and woodcutters are famed for their rude drunkenness by some, and for their egalitarian impulses by others. In either case, woodcutters created cultural forms that diverged from dominant European norms and values. Historian Mavis Campbell pithily summarizes accounts of the buccaneers written by two different men, Captain Nathaniel Uring and William Dampier, a privateer who himself was hired initially as a servant to work for a woodcutter before traveling around the world and writing his famous memoir. Campbell describes the woodcutters as a "hard-drinking, hard-swearing lot living in miserable conditions" (Campbell 2011, 112).[7] Uring's colorful passage depicting life in logwood camps has often been quoted and elaborates on the character and interests of these hewers of wood:

> generally a rude drunken Crew, some of which have been Pirates, and most of them Sailors; their chief Delight is in drinking and when broach a Quarter Cask of a Hogshead of Wine, they seldom stir from it while there is a drop left: It is the same thing they open a Hogshead of Bottle Ale or Cyder, keeping at it sometimes a Week together, drinking till they fall asleep; and as soon as they awake, at it again; without stirring off the Place. Rum Punch is their general Drink, which they'll sometimes sit several Days at also; they do most Work when they have no strong Drink, for while the Liquor is moving they don't care to leave it. (Uring 1726, 355–356)

Woodcutters' interests in activities other than work is supported by Dampier's comment that these men "thought it a dry Business to toil at Cutting Wood" (cited in Campbell 2011, 105). These commentaries describe the woodcutters as privileging sociality and pleasure over hard work and the pursuit of money. Woodcutters wanted to work on their own time, in their own way. These orientations remain a hallmark of life in rural Belize. As Linebaugh and Rediker suggest, communities like these reject the British Calvinist Protestant focus on self-restraint, delayed gratification and hard work that served as the cultural backbone for the development of capitalism, and that was gaining a foothold elsewhere in the late eighteenth century

Not only did they reject dominant British values, they also rejected Spanish and British colonial law. They directly defied international treaties governing land use. Britain's treaty with Spain allowed Britons only to cut wood, and specifically forbade them from permanently occupying or owning land in the territory. Yet the leading woodcutters both occupied and laid claim to these lands. However, the precarity of the wood-cutters' land claims made it very difficult for individual men to become wealthy through landholding.

The logwood cutters also developed relatively egalitarian political arrangements, in contrast to the increasingly hierarchical political formations capitalism was fostering elsewhere. As Mavis Campbell notes in her comprehensive history of early Belize, many of the buccaneers had been sailors, soldiers, and indentured servants "all from rigidly hierarchical systems" (Campbell 2011, 120) that they may well have wanted to leave behind. The earliest clear written record of their political arrangements dates from the 1760s, when a group of the settlers who called themselves the "Principal Inhabitants" created "Publick Meetings" from which they elected magistrates and created policy. In these meetings, the Principal Inhabitants established a rule by which to determine property rights. Those individuals who claimed a location for cutting logwood and who built a hut to mark it were deemed the owners of that property, or "works." The initial agreement among the logwood cutters ensured equity in access to resources, allowing any one cutter to "own" no more than two works in any given river, with each work limited to being only two thousand yards along the bank. At this point, these Principal Inhabitants did not include all the people living in Belize at the time, and specifically excluded women and enslaved people. Nonetheless, their efforts to ensure that no individual could amass property have been hailed by scholars and others for their relative egalitarianism at the time. This commitment to democracy and equality while revolutionary and utopian in some ways, also already codified difference and power in a way that would come to serve as the foundation for the class and color hierarchy that organizes everyday life in Belize today.

Race, Class, and Color in the Early Days of Logwood Cutting

These logwood settlements were taking hold as sugar production, plantation slavery, and the trade of enslaved Africans by free Europeans reigned across the Caribbean in the sixteenth through eighteenth centuries. Wealthy British men established plantations on Jamaica, Barbados and other islands in order to amass fortunes, often living in England and only visiting their holdings in the Caribbean occasionally. Along the coasts of Belize, a different sort of population was settling. The "motley crews" who found themselves on these shores were from early on a hardscrabble mix of poor white mostly English-speaking men: soldiers, sailors, and indentured servants (Campbell 2011), sometimes coming from other parts of the Caribbean to try their luck here, sometimes coming from Britain. There were also very few British women in the settlement. Enslaved people of African descent were present in Belize from early on, though in small numbers.

As logwood became more valuable, wealthier cutters began to import enslaved Africans and people of African descent from Jamaica to assist with the work. The first written record of slave-owning in British Honduras dates to 1724 (Bolland 1977; Campbell 2011); and in 1745, enslaved people of African descent outnumbered the free—120 to 50 (Campbell 2011, 297).[8] The number of people living in woodcutting camps throughout the 1700s fluctuated, but likely included slaves. At the same time, it is likely that free people of African descent, individuals who would be classified as "free-colored" (who appeared to have both European and African ancestry) or "free-black" (who appeared to have only African ancestry) were also among these groups of early woodcutters.[9] In the camps along the Belize River and in the lagoons where logwood could be cut, enslaved African man and woman, indentured European servant, and head English woodcutter mingled close together and a population of mixed English- and African-descent—some enslaved, some free—began to grow. This intimate mingling co-existed with the physical and symbolic violence of slavery, and British colonial law that codified that violence. White British men enslaved, raped, and impregnated African-descended women, sometimes then living with them and their children, sometimes not; sometimes freeing the women and their children, sometimes not. Enslaved African-descended people were always potentially subject to violent treatment by their owners, and were denied the rights, and the humanity, afforded free people.

This small population of people with different and violently unequal racial, color, and class positionalities began to create a common sustenance-oriented culture that deeply enmeshed them with the more than human found here. Using cultural heritages from both Africa and Britain, these displaced peoples re-rooted themselves here in the Americas (see Glave 2010; Mintz and Price 1992; Sheller 2003a), and together also created a common language, Belizean Kriol.

From the outset, social position and place intersected significantly in Belize. As noted above, wealthy British men were not typically attracted to this part of the Americas, so the place itself assembled British and others who were relatively less elite. The wealthier and higher-status Englishmen who did make their way to these shores set up their camps on the Cayes, away from Belize's swamps. Less elite white and both free and enslaved people of African descent were more numerous in the inland waterways of the territory, where they were engaged in the hard labor of cutting logwood and sending it down the waterways to the coast. Britishness and being economically elite began to be associated with the Cayes, and being of African descent and or poorer began to be associated with Belize's interior lowlands. Indeed, St. George's Caye where the wealthiest individuals lived in the 1700s is a fancy resort today and still home to some of the wealthiest Belizeans.[10]

The Rise of Mahogany

By the late eighteenth century, mahogany began to supersede logwood as the most valuable export out of Belize. Logwood was still cut and sold, and was an important part of the colonial economy, and indeed, the demand for logwood continues up to the present day.[11] But mahogany began to be a highly valuable wood in England early in the eighteenth century, as it became the prized furniture wood for upper- and middle-class consumers there (Anderson 2012).

This shift had a profound impact on the logwood-cutting society that had developed here. Mahogany liveliness and ecological relations precluded individual woodcutters from making a living cutting these trees. Unlike logwood which grows in dense shrubby clusters along waterways, mahogany trees grow singly and widely dispersed in higher, drier lands that Belizeans call "high ridge," which can occur in extensive tracts or smaller copses, or "cayes," in the pine-savannah, and which are often far from waterways. Furthermore, the tree trunks most coveted by traders were enormous (trees grow up to 120 feet high, with diameters that can be greater than 10 feet). A group of men was required to cut trees this large and this widely scattered, and to transport them to rivers deep enough to float them. This need for additional labor generated a demand for slaves and the number of slaves in the colony grew rapidly as the mahogany industry expanded. This expansion was likewise accompanied by increasing inequality: only men with capital could enter into mahogany cutting. The color-class hierarchy that had begun to emerge with the logwood industry became much further entrenched as the mahogany industry grew.

Continuing tensions and skirmishes between England and Spain meant that woodcutters living in the Bay of Honduras were never settled comfortably until the end of the eighteenth century. Spanish raids routinely removed woodcutters from Belize, and Britain's other Central American claim, the Mosquito Shore, or the northeastern coast of present-day Nicaragua.[12] A significant shift occurred with a 1786 treaty in which Britain agreed to give up the Mosquito Shore in exchange for a (slightly) more secure hold of territory in Belize. Over the course of the next four years, nearly two thousand Mosquito Shore residents left for what was then a very sparsely populated Belize, where only a few hundred lived beforehand. These evacuees included a large number of enslaved people of African descent, poorer whites and poor free-colored and free-black and smaller numbers of white and free-colored slave-owners.[13] By this time, the opportunities afforded and denied free-colored and free-blacks by colonial law were increasingly codified. Although the details varied by colonial power, slaves were property of their masters and had extremely limited rights, free-blacks had lower status and less rights than free-colored; and free-colored less than any white, whether wealthy or poor. Throughout the Caribbean, the free-colored, who occasionally inherited or were able to accumulate wealth and sometimes owned slaves, occupied an ambiguous status that threatened white slave-holding hegemony.

In Belize, the increasingly elite "Principal Inhabitants and Merchants," as the wealthy white Baymen called themselves, were concerned about the evacuation of the Mosquito Shore. Many of the evacuees who were poor whites or free people of color would need to be resettled, threatening the wealthy Baymen's potential ownership of mahogany works. Particularly worrisome to these white British Baymen were the free-colored who owned slaves and who at times claimed equal status to the Baymen. The Principal Inhabitants quickly established rules for how land could be allocated to all people now living in Belize, stipulating the amount of capital, or number of slaves, necessary to qualify to occupy works. The restrictions were greatest for free-blacks, slightly less for free-colored, and least for whites. The Baymen, lauded for their initially revolutionary egalitarian and self-governing Public Meeting in the era of logwood began to codify a class and color hierarchy in Belize that looked very similar to the rest of the Caribbean, ensuring that lands where mahogany was plentiful were kept out of the hands of poor and nonwhite evacuees.

Race, Color, Class, and Place in the Mahogany Era

What impact did these new land-use regulations and influx of people from the Mosquito Shore have on the settlements in the lagoons at Crooked Tree and along Spanish Creek? A map drawn in 1787 (see figure 2.2) indicated that even before the evacuation from the Mosquito Shore, the area had already been logged out of mahogany.[14]

The dearth of mahogany and the likely presence of logwood cutters, rather than wealthy mahogany works owners, meant that these areas may have been open to evacuee settlement. A number of the evacuated free-colored had surnames today associated with these communities. One of the free-colored evacuee surnames was Crawford, a name associated with Crooked Tree, and another name that appears on the post-resettlement census of 1790 is Thomas Bonner, a name associated with Spanish Creek and Lemonal, and an ancestor of my husband and children. Furthermore, to this day there is a sense of historical connection between this area and northeastern Nicaragua, or what Belizean Creole people call "Bluefields," in reference to its main population center.[15] Older individuals share stories of ancestors who hail from Bluefields, or who traveled back and forth from there in the past.

Poorer whites and free people of color, along with the relatively small number of slaves they owned, whether from the Mosquito Shore or not, could make livings on the sides of shallow waterways.[16] When mahogany replaced logwood as the most valuable export and the local governing elite limited landownership possibilities, relationships between race, class, and place likewise shifted. Phenotypically white English people, who had little melanin in their skin, narrow noses, and straight hair lived in houses in the Cayes and Belize City, but had large landholdings upriver that stretched away from the riverbank into the bush, where mahogany could be found. Their phenotypically black African slaves, who had more melanin in the skin, broader and shorter noses, and curlier hair, spent most of each year in camps on those extensive upriver landholdings, in mahogany gangs cutting and hauling huge mahogany trees. Poorer whites and free-colored with in-between phenotypes found ways to continue to live along shallow lagoons and waterways, by cutting and trading logwood, engaging in other extractive activities and cultivating, fishing, and hunting.

Colonial discourse encouraged these linkages between race and place. For example, Captain G. Henderson of His Majesty's 5th West India Regiment visited British Honduras and published a full description of the colony in 1809, shortly after the British slave trade was abolished (Henderson 1809, 49). In his careful and elaborate description of mahogany industry operations, Henderson uses language that racializes "the huntsman," the "negro slave" designated to find mahogany trees. He finds remarkable the huntsman's innate capacity (rather than creative thought and intelligence) to always find his tree. Later,

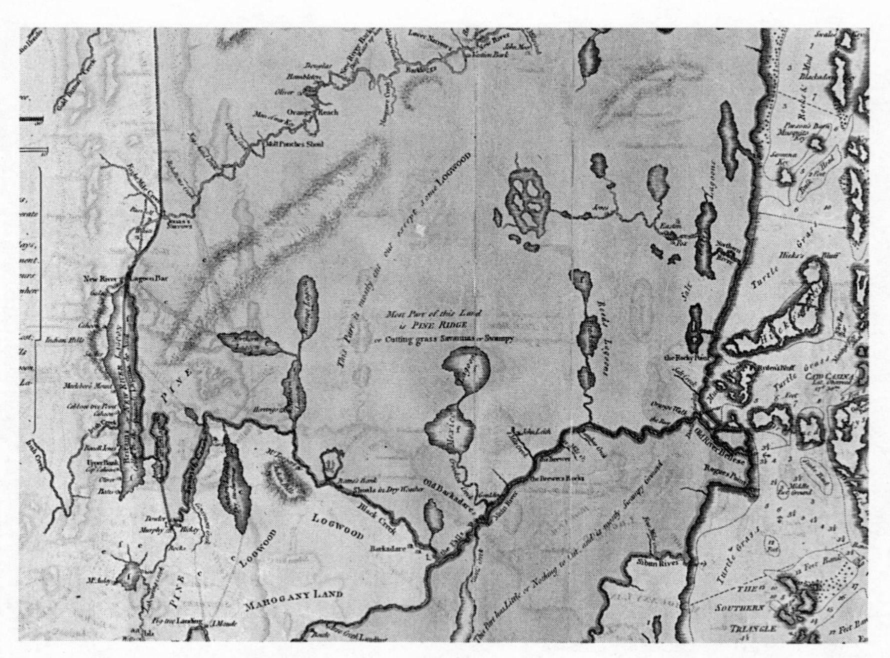

Figure 2.2 Close-up of 1787 map of the Yucatan, indicating areas cleared of mahogany. *A map of a part of Yucatan, or of that part of the eastern shore within the Bay of Honduras alloted to Great Britain for the cutting of logwood, in consequence of the Convention signed with Spain on the 14th July 1786*, by William Faden, London, 1787. Accessed September 28, 2017, from the Library of Congress, https://www.loc.gov/item/gm70000406/.

Henderson comments on the utter comfort of the huntsman in these tangled tropical forests (see Johnson 2003). In both of these instances, the animality of the huntsman is emphasized, contributing to the association of African-descendedness and the jungle.

This purported capacity for blacks to do forest labor, of course, served to justify their enslavement. The *Honduras Almanack*, a colonial publication designed to appeal to potential travelers to and investors in British Honduras, and published just prior to the abolition of slavery, suggests the association of "black" and "colored" persons with swamps and woods, but in a less laudatory fashion. The concern of the *Almanack*'s "white" British authors is that when "black" and "colored" people are free, they are worthless to the aims of the colony. The authors describe a typical free-black/free-colored as

> equally ignorant of the value of time and responsibility for the improvement of (the country), is not alive to the duty of industry: clothes, he requires none beyond a shirt and trousers, and a small quantity of powder and shot, a few

> hooks and line, will in half an hour, furnish him with sufficient to support his family for a day or two. The want of education leaves him void of inducements; and he seems happiest when settled in a swamp and surrounded by mosquitoes. Hence free labour is exorbitant . . . very few industrious men can be found. (*Honduras Almanack* 1830, 7–8)

These two pieces of colonial writing articulate dominant British ideas about who belonged where doing what and show how phenotypic characteristics assembled with particular places and with conditions of freedom or enslavement. Numerous propositions about blackness were encoded: (1) innate capacity and ability in wild places, as Henderson so admired in the African-descended enslaved person who worked as a huntsman, (2) an innate desire to be in the forest, to work as woodcutters, and the association of the bush, uncleared wild, often forested areas, with blackness, (3) the association of blackness with laziness and the inability to productively transform landscapes for industrial agriculture, and (4) that people of African descent were most suited to be fishing in mosquito-filled swamps. These conceptions also implicitly encoded the opposite for whiteness—that white intention, ambition, and industriousness precluded living in swamps, and that whiteness was best suited to transforming landscapes.

Assumptions about race and landscape were not all that were encoded in the commentary. It also contained implicit evaluations of worthy and unworthy modes of living in and using landscapes. Colonial commentators aimed to identify ways that profits could be made to grow the colony—timber production, large-scale agriculture, and so on. They chose *not* to see sustenance-oriented plantations, cattle husbandry, fishing and hunting that prevailed in the small settlements along river banks as productive. These small-scale socio-ecologies, often tended by free black and brown people and also poor whites, constituted ways of making a living in these places, but were seen as wasteful and unproductive by the colonial elite.

It is not coincidental that this second commentary was produced as the abolitionist movement was gathering steam, and Britain was on the verge of emancipating enslaved peoples. In Belize, abolition occurred at the height of slavery's utility, when the mahogany industry was at its peak. Wealthier mahogany cutters and traders were concerned they would lose the labor force they needed to keep their business growing. They worked with the British Colonial Office to put into place land and labor laws to prevent newly freed slaves from seeking work elsewhere, or from earning a full living planting rich agricultural lands (Bolland 1977). These laws were supported by this colonial discourse, which consolidated the association of blackness with swamps, woodcutting, and the inability to develop agriculture.

Crooked Tree and Lemonal, 1832–1850

A rich description of life in Crooked Tree and Spanish Creek in the mid-nineteenth century illustrates the historical themes I have outlined. English Baptist missionary Frederick Crowe who helped establish the Baptist church in Belize describes in detail his travels in British Honduras between 1832 and 1850, as abolition took place (Crowe 1850). Along with recounting his own experiences, Crowe also includes letters written to him by his assistants, one of whom, Mr. Henderson, was charged with expanding the geographical scope of the Baptist church from Belize City out into the districts. Crowe's account reveals the details of how people became entangled with the natural world in these places and how racial hierarchy shaped life here, as well as the role of the Baptist church in the history of this place.

Crowe recounts a letter he received from Henderson that describes the work of "Brother Warner" who had started out his tenure in Belize as an impoverished and sick English sailor in the 1830s. Warner had been "rescued" from "debauchery" by a Scottish fisherman who made a living on Turneffe Caye raising pigs and collecting turtles that he occasionally brought to Belize City to sell. Warner became an acolyte of the Baptist church and started to work for Henderson. Warner established a "station" (or church and school) at Spanish Creek, which in 1843 had twenty young students, indicating a significant settlement at what is now known as Lemonal village (Crowe 1850, 368). From here, Warner set up a station at Crooked Tree, and I quote from Crowe's recounting of a letter from Henderson, dated May 9, 1844:

> John Warner had not long before been removed from Spanish Creek and stationed at Crooked Tree, a place which he had just returned from visiting, and which he describes as a collection of about 20 houses belonging to mahogany and logwood cutters, situated on an island in the midst of a beautiful lake called Northern Lagoon. It is open to the sea-breeze, and was evidently the site of an ancient Indian settlement. It is about 50 miles NW of Belize, surrounded by extensive plains of open pine ridge, and is reached by ascending the river Belize, which has some falls rather dangerous to pass, and then paddling up a tributary stream called Black Creek. At that time it was the residence of William Tillett, Esq., whose family alone consisted of 15 children, besides numerous dependents. This gentleman, brother to the generous friend of Baker's, was also an extensive proprietor of land, and favoured the formation of a station on his property. As was commonly the case in his excursions, Mr. Henderson was on this occasion requested to marry several people. This is often the first outward step towards a profession of religion, the parties frequently making a tardy reparation for their neglect in this matter, sometimes in the presence of their children. Being requested

> to lay down a plan for a future town and to give it a name, he did so—calling it *Tilletton*. He says—"At Tilletton two of our members already have their residence, two more have houses in which they occasionally reside, and four from Spanish Creek about ten miles off (where Br. Warner had already been blessed) speak of removing thither. Two families, the elder branches of which had been seriously impressed under the means of grace at Baker's, have lately taken up residence at Tilletton." (Crowe 1850, 371)

Henderson writes to Crowe in 1846, two years later, noting that several hundred sinners "were converted to God" (Crowe 1850, 428), indicating that the settlement was well populated. Crowe's description shows just how far "off the main" these locations are. It takes a long time, and is not an easy trip, to reach these places. It is a two- to three-day paddle at least, to reach "Tilletton" from Belize City; and this remained the case until late into the twentieth century. Some people lived here full time, some occasionally resided here, and people would move residences from one settlement to another easily.

The people who lived here cut mahogany and logwood for a living. The wealthiest men, like William Tillett, had slaves or could pay laborers to travel to the high bush to fell and transport mahogany. But residents also sustained themselves in a variety of other ways. Elsewhere Crowe describes Warner's travails when he was given permission to cut plantain on William Tillett's plantation, which from the description, is located today where many Crooked Tree agricultural plots and cattle ranches are located, on the western side of Western Lagoon, a good two- to three-hour walk or forty-minute horseback ride from Crooked Tree (see Map 1.2). He writes, "He [Warner] had to traverse the lake and an extensive swamp for a considerable distance, wading through the water, his clothes and skin lacerated with cutting grass, his face and hands stung by mosquitoes, etc., and in danger of putting his foot upon the scaly back of an alligator at every step he took. He had of course to return by the same way with a load of plantain upon his back. These laborious expeditions were generally followed by a fit of fever and ague" (Crowe 1850, 380).

At another point in Crowe's text, a preacher expresses thanks and praise for the flourishing of plantations at Crooked Tree. Recording a letter from Henderson again, he notes: "Tilletton is in 'great joy!' The blessing of the Lord is on the very ground about them. Their plantations flourish" (Crowe 1850, 460). And in yet another passage, we are introduced to Alexander Kerr, who had been brought into the church at Tilletton and then was formally baptized in Belize City.

> One was Mr. Alexander Kerr, a Belize Creole, who had been converted through the instrumentality of Brother Warner, of Tilletton. He was a hunter of deer and antelopes in the pine ridges of the old river, and excelled in making moccasins, a kind of buckskin boots, from their hides, which he prepared with his own hands. He had, in his youth, enjoyed the advantage of some

> education at the free-school, and seemed likely, both by natural gifts and a sanctified disposition, to be an instrument of usefulness in evangelizing the country. He was then in the full vigour of manhood, and at once a husband and a father. (Crowe 1850, 426)

Alexander Kerr was a hunter and a craftsman, and it is likely that he also labored as a woodcutter at times. Moccasins loom large in any conversation with rural Belizeans about times past. Until very recently there were always people who were famed moccasin makers in Crooked Tree and Lemonal; moccasins could be made from any leather—but the best ones were made with deer hides. Alexander Kerr is evidence of the very long history of this particular entanglement with the more than human. These passages reveal that people sustained themselves using a wide range of interactions with the more than human in the area. What rural Belizeans call "kech and kill," making a living through a variety of activities, was well established in this place by 1850.

Just as Crowe's account provides insight into the socio-ecological entanglements that characterized this part of Belize in the mid-1800s, his writing, in combination with information from censuses conducted just before and after the abolition of slavery, reveal how shifting race, color, and class hierarchies became assembled with different places in Belize. His characterization of the population of Belize as a whole is worth reading:

> Mixed and varied as the inhabitants of Central America must appear to the reader, no part of it, and perhaps no place of the size anywhere, presents so great a medley as Belize, where are to be found representatives of each of the races, pure and mixed, already described and almost of each tribe already mentioned [European, African, a wide variety of groups native to other parts of Central America], together with Conges, Nangoes, Mongolas, Ashantees, Eboes and other African tribes. Among the Belize Creoles are included every shade of colour and admixture; the Mulatto and the Sambo, as understood in the West Indies, and all the other degrees, such as Quadroosn, Mustee, &tc. &tc. &tc. There are also Creoles from Jamaica and the Bahamas, as well as from the French, Spanish, Dutch and Danish Islands, New Englanders, South Americans, and a great variety of Europeans—among which the Scotch and English predominate—besides adventurers from almost every part of the world, and some few Jews. (Crowe 1850, 50–51)

In this passage, many different types of people, with varied origins, are recognized. Racialization has not fully consolidated and submerged the differences among those who will become classified as "black," "colored," or "white" on later censuses. Countries and language groups still differentiate individuals. Yet, even as Crowe delineates these differences in this general description of the colony, when he describes his travels through Belize, and the history of the Baptist church there, he categorizes the people living in Crooked Tree and Spanish Creek as Belize Creole, but the Baptist church visitors from England are usually not identified. The whiteness of being English is unmarked in this text and is instead taken as a default.

Racialization and the Case of the Tillett Brothers

Key figures in the establishment of the Baptist church in Crooked Tree were the Tillett brothers: William and George, both described as "Creole" rather than white, born in Belize and of mixed race heritage, but also both listed as "Esq." indicating that they were men of high socioeconomic status. George Tillett had a house and land in Belize City on the river front, as well as one at Baker's Bank, on the Belize River. Newspaper listings in the mid-1800s include him among the leading mahogany cutters in the settlement. While not as elite as George, William Tillett has a large property at Crooked Tree. In Crowe's account, William Tillett had fifteen children, his wife, and numerous dependents. What makes all of this particularly interesting is how these brothers are entered first on the census of 1832, two years before the abolition of slavery and then again, on the census of 1835, one year after abolition. In the 1832 census, William is listed as a "coloured man" living with a "coloured woman" and their eight "coloured children." According to this census, he also owned five slaves: two men in their thirties, two small children, and a woman in her thirties. Three years later, by 1835, William, like his brother George, was listed as a "white man" married to a "white woman," with eight "white children" living with them, but now having in their care thirteen apprentice-laborers, just enough to work mahogany. These two Tillett brothers are the sons of English sea captain William Tillett who arrived in Belize in 1784, and a woman, Elizabeth. Elizabeth in turn was the daughter of English sea captain William White, and a woman, Mary, who is described as "mixed," which most likely refers to her being "Miskito Indian." The Miskito, indigenous peoples native to the northeastern coast of modern-day Nicaragua, were thoroughly intermarried with people of African descent, suggesting that she was partially "black" as well as "Indian."[17]

As happened in other parts of the Caribbean, people whose embodiment was close to white indeed became white after the abolition of slavery, especially when they were economically well-off. Now that the distinction between enslaved and free was officially erased, being white served as a more significant identifier of status. But only wealthier lighter-skinned individuals more connected to Belize City and the Cayes, like George and William, could make these kinds of claims to the census takers. Other Belizean Creoles continued to be classified as "colored" and not white. The vast majority of the individuals living in these rural communities, even if embodied in a way that almost appeared white, remained "colored" in the official records. Living in rural out-of-the-way places, and not being well-to-do associated them with being non-elite and nonwhite, like Alexander Kerr, the Belizean Creole man described above who was a deer-hunter, moccasin-maker and church member. Interestingly, Kerr was also married to one of William Tillett's daughters. That Kerr was a suitable match for William Tillett's daughter may or may not have been associated with his being a "clear" or light-skinned Belizean Creole. Crowe never describes his embodiment in any more detail than "Creole." His eligibility certainly also seems to be connected to his respectability; he worked as a preacher for the Baptist church.

Crowe's text raises another critical point about the realities of post-emancipation societies that sheds additional light on processes of racialization in the decades after abolition. Crowe recounts a letter from Henderson in which he describes how people were baptized in the Crooked Tree lagoon waters by the meeting house: "The lame ones walked in with their *slaves*, or supported by the brethren" (Crowe 1850, 439; emphasis added). This letter was from 1848, fourteen years after the abolition of slavery, eight years after the last slaves would have ended their apprentice period.[18] Why does Henderson refer to these people as slaves? Could it be that slavery de facto continued on, even if it was technically illegal? Were individuals who had been enslaved never set free? Had they been freed but then coerced or somehow made to continue working as servants for their prior owners? Although it is impossible to tell how widespread this was, that it was so nonchalantly reported by a chronicler who elsewhere rails against the institution of slavery suggests that it might well have been common. Given that people who were darker-skinned were more likely to have been slaves, the persistence of the designation "slave" for at least fourteen years after abolition further strengthened a close association of blackness with not being free and economic disenfranchisement.

People from this part of Belize today find amusement in some of the details of the historical assemblages of race and color here. Stories of lighter-skinned people "skahning" (scorning) darker-skinned ones are told with an amused chuckle. A story from Crooked Tree's past tells of a clear-skinned woman who wanted nothing to do with black individuals. If a dark-skinned person visited her house and asked for a cup of water, the visitor would be offered a tin cup, not a glass. They might even have been given the cup with a "kiss-kiss," a palm stem that forms natural tongs, to ensure that the light-skinned server would not have to physically touch the black visitor.

According to oral history in Crooked Tree, the Tilletts and the Rhaburns were the largest landowning families, suggesting that William Tillett may have been one of the (if not the) wealthiest resident in Crooked Tree in the mid-1800s. To this day, in Crooked Tree, many, but certainly not all, individuals with the surname Tillett have light skin; and the village of Crooked Tree in particular is famous for its residents being light-skinned. So, while living along swampy lowlands disassembles whiteness from a person, this particular location of

Crooked Tree is also disassembled from blackness. As racial categories gathered more associations and became more sedimented, so too did the entanglement of an emerging Belizean Creoleness—neither white English nor black African—with these logwood-rich waterways and their banks.

Crowe's descriptions also support my contention that the people who chose to live here in Crooked Tree (Tilletton), Lemonal (Spanish Creek), Backlanding, Revenge, and other settlements in the area also created ways of living economically, ecologically, and in this case, even spiritually, of their own making. Individuals here made plantations, hunted deer, made moccasins, and also cut wood to sell. And when they engaged the presence of the English Baptist church, they did so on their own terms. Toward the end of Crowe's account, he reports on tension over land rights between the Baptist Mission Society and William Tillett's estate (Tillett having died in 1848 of "fever"), which irked the people of Crooked Tree. At the same time, Alexander Kerr, the moccasin maker, had established himself as an independent Baptist preacher and had moved to Backlanding, yet another settlement in the area (see Map 1.2).[19] Because the residents of Crooked Tree were displeased with the Baptist Mission Society, Crowe reports, a great many of them moved to live in Backlanding to be served by Kerr.

Although we cannot be sure about what motivated people to move, that they did move so easily is significant. The picture that emerges of life in this part of Belize at this time is one of great fluidity and flexibility. People knew the area well and moved comfortably between the various settlements, to take advantage of whatever possibilities might be available. Although undoubtedly there were deeply pious people among the residents of Crooked Tree, I suspect that the great interest in following the Baptist missionary stemmed equally from the desire to benefit from the educational opportunities for children offered by the church.

It is significant that the Baptist church was such a critical force at this time. Assad Shoman, in his valuable history of Belize, argues that the Anglican and Baptist churches served as a means by which European values were disseminated into the lower classes in Belize (Shoman 1994, 147–148). While to some extent this may have been true, what happened here may have been more like what Karen Olwig describes for Nevis (2002). The presence of the church did not mean that the European values of respectability were simply being fully assimilated by rural Belizeans, but rather a more complex process was occurring. Some values were incorporated into the growing amalgam that was Creole culture, some were rejected. The turbulent history of the church in Crooked Tree is plain evidence that a hegemony of middle-class European values was not easily (if ever) established. Warner's house was set on fire (Crowe 1850, 380), and people abandoned the Baptist church and its foreign-born preacher for the ministrations of Alexander Kerr, born and raised in Crooked Tree (Crowe 1850, 484–488). But only two years into his service, Kerr complained of losing a large part of his congregation because they became too interested in the ways of the world, rather than the ways of the spirit: "the influence of the world having been suffered to assume ascendancy over the minds of some members" (Crowe 1850, 494). That a relatively wealthy, light-skinned Creole man who was designated "white" on the first post-abolition census was the church's most ardent supporter suggests that church identification had a symbolic association with whiteness and higher economic status. For William Tillett, wanting to claim whiteness and being a part of the church fit together as part of a construction of an elite identity. The church may have served to focus debate within the growing Creole community about what constituted Creole culture, and the nature of the relationship between Creole culture and colonial cultural politics. For others, this symbolic association of whiteness and eliteness may have been experienced as exclusionary, and thus made the church an unattractive and uninteresting institution.

In sum, at the height of mahogany extraction in Belize, the racial and ecological orders solidified and the most elite white mahogany extractors and traders lived in Belize City or on St. George's Caye, with populations of slaves working for them upriver. Meanwhile a not insignificant population of free black and brown people and less wealthy whites were settling along the waterways of the lower Belize River Valley. Laws, ecological conditions, and colonial discourse all convened to associate being Belize Creole (something between African- and English-descended) with these landscapes. The Belizean Creoles who lived here crafted socio-ecological economic relations on the margin of the extractive capitalism of the mahogany industry. Through re-rooting themselves in these lands, they crafted a culture and a language. Although these emerged out of the violence of slavery and reproduced the color hierarchy that justified the enslavement of black people, this culture and language also produced positive senses of belonging and identity that coalesced with a distance from whiteness and proximity to blackness.

From 1860 to the Mid-1900s: Timber, Chicle, and Peopling the Waterways

By the mid-nineteenth century, Belize's top mahogany cutters and merchants had emerged as what Nigel Bolland dubs a "forestocracy" (Bolland 1977). The territory changed status to a Crown colony in 1862. At the same time, the largest mahogany operations joined to form a new company that would come to dominate Belize's economy for almost one hundred years. The Belize Estate and Produce Company (BEC) owned a staggering half of all privately held land by 1880, 1.5 million acres, or one-fifth of the colony as a whole (Bolland and Shoman 1977, 82; Judd 1992, 73).

Although Belize's Crown colony status indicates a greater interest from Britain, BEC's dominating presence and its timber extraction orientation meant that Belize experienced very little infrastructural development until well into the twentieth century. Belize was traversable primarily by river and waterway until the 1940s. Crooked Tree, Lemonal, and other villages were not readily accessible by vehicle until the 1980s. Many people I spoke with in the 1990s and early 2000s recalled traveling by dory, or canoe, to Belize City in their youth.

Two important shifts took place as the twentieth century neared. First, this British colony began to orient itself economically more to the United States than across the Atlantic to England. Second, the extraction of natural resources expanded beyond mahogany. Logwood continued to be exported, but a host of other forest products became increasingly important. At the turn of the twentieth century, a new fad hit the streets of the United States: chewing gum. The William Wrigley company in Chicago dominated the gum market and demand skyrocketed for chicle, the sticky substance that was first used to make gum. Chicle is the boiled-down sap of the sapodilla tree, which grows abundantly in the forests of the American isthmus. "Chicleros" tap the sap, just as a rubber or maple tree might be tapped, collect it in buckets, and sell it. Many rural Belizeans, including my father-in-law, tapped chicle, or worked as chicleros as one among a suite of occupations they pursued. Robert Thurton, a wealthy British man who owned large tracts of land in northern Belize served as the agent for Wrigley in the early twentieth century, and tales abound of rural Creole men's experiences dealing with "Turton," as he was called in Belizean Kriol.[20] By the 1920s, mahogany, cedar, and chicle constituted 82 percent of total exports out of the colony, with the vast majority going to the United States.

Creole men then living in rural Belize often worked for periods of time for the BEC. Some had high-status jobs like running the company's commissary at Gallon Jug, which was the main camp of the company; others worked in the myriad positions that constituted forestry work in Belize, such as carpentry, cattle husbanding, woodcutting, and engine repair. BEC workers were paid via the notoriously exploitive "advance" system. First, workers were typically given about half their wages in overpriced company store goods (clothing and foodstuffs) and the other half in cash. Second, when laborers completed a work period of several weeks, they would often celebrate, borrowing against their future earnings from the company, thus placing themselves in long-term debt to their employer (Judd 1992, 179).

One other significant development at this time was the growing concern for the continued health of Belize's forests and animals. A forestry department was established in 1923 to improve the competitiveness of Belize's wood industry and to ensure that this industry would not destroy itself (Benya 1979; Hummel 1921). Colonial records from the early twentieth century also contain numerous reports and complaints about the depletion of game animals and fisheries (especially marine turtles). Spanish Creek (between Crooked Tree and Lemonal) figures centrally in concern about over-hunting and the eventual establishment of a closed season for deer in Belize (MP 4056 of 1915).

While the historical record is skimpy, birth registries and other descriptions of the area at the time suggest that people remained in these areas through generations. Indeed, most of the family names that are today associated with different villages and other smaller settlements in the lower Belize River Valley are found connected to these places from the earliest days of the nineteenth century. This depth of familial connection to these lands and waterways, to being planters, cattleman, cooks, woodcutters, laborers, teachers, and all other manner of occupations in these places, suggests an equally deep connection between identity and the more than human found here—the game animals, the different tree species, fish, turtles, the half-wild cattle, the fruits, suckers, and ground foods that people plant. Like Alexander Kerr, well known for making moccasins in the mid-nineteenth century, the Tilletts, Wades, Bonners, and Crawfords were all known as good "bushmen." Oral history of these places corroborates this picture.

A variety of historical sources suggest that a large population of Belizeans, and specifically people identified as Belizean Creole, were living along the rivers and waterways throughout the country (Fowler 1879). Conventional histories of Belize assume that Belize City was the stronghold of Creole culture. Relatedly, it has been assumed in discussions of Belize's history that gangs of men went to work at mahogany cutting camps, leaving behind their families, their women, and children in Belize City. While an elite Creole culture that appeared to emulate British middle- to upper-class culture may have been amalgamating in Belize City, and while poor black laborers were also likely creating their own version of Belizean Creole culture in the city, I contend that what it means to be Belizean Creole was being forged at other sites.

The Birth Records of Orange Walk District for 1913 reveal a great deal about this part of the colony at the turn of the century. Births from 1888 through 1913 are recorded here, and and many contain either surnames or place names associated with Lemonal, Crooked Tree, Backlanding, Revenge, Bermudian Landing, and other neighboring villages and settlements.[21] The records show that people lived in a wide array of settlements throughout this area, and that there were women (bearing children) in many, if not all, logging camps and settlements. In the space of ten years, two or more births (up to thirty for some locations) were recorded for thirty-five camps or settlements in the late 1800s. Men did indeed take their women and children to camp with them, and women worked in these camps as washerwomen and cooks, and in other kinds of occupations.

Since many couples had more than one child in the space of ten years, the records also show the mobility of people in these areas. Some couples moved every year, sometimes living in logging camps, sometimes in villages, having babies as often in a logging camp as in a village. Others stayed in the same place for all that time, bearing numerous children in one location. The records also indicate (if known) the father's occupation, occasionally the mother's occupation, and, if the information was reported by someone other than the mother or the mother's husband, the occupation of the reporter. The majority of men worked as "laborers," and occasionally the kind of labor was specified as mahogany cutter or woodcutter. But large numbers of men are also identified as logwood cutters, planters, and cattlemen. A fair number of women are listed as washerwomen, seamstresses, domestic help, and housekeepers. A few men are listed as "Captains" of labor, mahogany, or logwood, or as "Contractors." A varied set of occupations thus characterizes these lagoon- and water-crossed lands at the turn of the century, a variation and multiplicity that is still present today. A comparison between specific individuals' job classifications in these documents with oral history and family memories of how these individuals made a living (who are the parents and grandparents of people still living in Crooked Tree, Lemonal, and Bermudian Landing, for instance) makes the picture even more complex. Men who are classified in the birth records as "laborers" might also have been planters, preachers, and teachers at the same time. Likewise, some men whose occupations are listed as "Captain," are known to have spent much of their time as young men raising cattle in Crooked Tree. On the other hand, some people are listed as planters who also worked as logwood cutters, or as laborers in the mahogany camps. For example, Elijah Bonner is listed as a "laborer" on a 1908 birth record of one of his daughters, but during this time he was also a teacher in the Baptist Church School in Crooked Tree.[22] Of course, each record is a snapshot in time, of when a birth occurred, where the mother was living, what the mother and father did to make livings at that moment. And perhaps that is the crux. People moved around this landscape frequently and engaged in different activities to make a living, and most important, people did not just work at one logging camp and then head to town during the off-season.

Although the reign of logwood had long ended and mahogany's banner years had passed, at the turn of the twentieth century people living in this area were still cutting logwood and small amounts of mahogany either independently or as wage laborers for the BEC and other companies. There are ample applications for logwood licenses in the colonial records of the early twentieth century for this area, most of which appear to have been granted as a matter of routine. Rural Creole people also planted, fished, and hunted, providing both material goods and cash income for themselves through their more-than-human engagements. By this time, merchants were selling their wares up the rivers and their tributaries (Leslie 1987), so rural people could purchase imported goods to supplement what they were able to provide for themselves. Baptist missionaries, preachers, and school teachers were assigned to Crooked Tree, Lemonal, and Backlanding throughout this time (Cleghorn 1939).

Perhaps it does not take a huge leap at this point to imagine that being Creole means being able to move around in this bush, to know the places where one can live and work, to know how to sustain oneself with the plants, animals, water, soils, and wind found in each place. Becoming Creole means knowing how to take advantage of any new opportunity that presents itself: from cutting logwood to running a commissary, making deer moccasins, or bleeding chicle. During the course of the nineteenth century, British Honduras became home to three more ethnic groups. These groups not only contributed to defining what Belizean Creole means but also further solidified the association of Belizean Creoles with particular parts of the country, and especially the area of Belize that is the focus of this book.

Belize's Developing Ethnic and Racial Matrix

In the early 1800s, increasing numbers of Black Carib, or Garifuna, people moved into Belize, fleeing persecution in St. Vincent and Roatan, Honduras.[23] A large number "completed" the migration process in 1832, shortly after they had backed the wrong party in elections in Honduras. These people were the descendants of mutineer Africans and the Carib Indians whom these mutineers encountered and then joined on the island of St. Vincent. In terms of understanding the social matrix in Belize, their dark skin color, African-indigenous language, and freedom made them fall outside of the emerging Belizean racial logic. Their blackness placed them at the bottom of the social order, as did their language and customs. On the other hand, they were free and were reputed to be hard workers—the epithet of lazy was not accorded them in the colonial descriptions of different ethnic groups as it was for Creole peoples (Johnson 2005; see also *Honduras Almanack* 1830, 1839; Morris 1883). In this way, the Garifuna gave the emerging category of Belizean Creole a higher status (by virtue of color and not being indigenous), at the same time that they threatened the Creoles' preeminent status as the backbone of the colony. This uncomfortable relationship between Creole and Garifuna continues to this day. A high proportion of Belize's most illustrious and highly educated citizens are Garifuna—teachers, doctors, government officials—despite the fact that Garifuna make up only a small percentage of the population. Yet poor Garifuna are likely to be treated as though they are at the very bottom of the social barrel.[24]

The Yucatec Caste Wars of the late nineteenth century prompted the migration of Yucatec Maya from Mexico into Belize, increasing the numbers of Mestizo and Maya in the country.[25] At the outset, these people were likely thought of very negatively as some sort of difficult cross between what the British settlers saw as marauding bands of Indians and the despised Spanish who claimed this turf as their own. On the other hand, the birth records (births from 1888 to 1913) for the Orange Walk District for 1913 indicate that people with Yucatec Maya surnames were sometimes living and working side by side with Belize Creole people. But because the territory was British and because it had been wrested from the Spanish, those Belizean Creoles who felt a strong affiliation with Britain (particularly light-skinned individuals), likely felt themselves superior to the Maya/Mestizo. The relations between Belizean Creoles and newly incoming Maya would also have been shaped by Belizean Creole's prior experiences with groups of Maya who periodically raided British encampments and settlements. As late as 1867, two logwood works owners registered a complaint with the Colonial Office about how their works at Indian Church in New River were under threat of attack by "wild Indians," who had already attacked them once with bows and arrows (Old Records 1867). Thus, the Belize Creole way of thinking about and dealing with Maya people involved both a wariness and a feeling of superiority. Most Yucatec however, lived in the more northern parts of the country, just as Kekchi and Mopan lived in the West and South of Belize. This spatialization intensified the Belizean Creole association with the lower Belize River Valley and Black and Spanish Creeks and the lagoons at Crooked Tree. When a Mayan or "Spanish" person married into one of these communities, their ethnic identity became subsumed into Creole.

The third group to arrive in Belize in the late nineteenth century was East Indians. They were brought as indentured servants to the Caribbean as a labor supply after emancipation, but many remained after their contracts expired. Three thousand came to Belize in the late 1800s, and some villages in today are still primarily East Indian.

This increasingly complex ethnic matrix strengthened the shared sense of ethnic identity among people who considered themselves Belizean Creoles, and the spatial separation of these groups aligned with colonial discourse about race and about what kinds of bodies belonged where. Belizean Creoles continued to be seen by colonial elite as inherently averse to agriculture, and therefore, by logical extension, only "useful" in the woodcutting industry. Yet all the while, rural Belizean Creole people planted a variety of crops and raised cattle and other animals in addition to hunting, fishing, cutting wood and using the more than human in a panoply of ways to sustain themselves.

Racialization processes distinguished between Belize's ethnic groups, and contributed to associating Belizean Creole people with the lands and waters of the lower Belize River Valley. Yet these processes also operated *within* the population identified as Belizean Creole, skin-color distinctions, or colorism, is a prominent feature of everyday life among both rural and urban Creole Belizeans.[26] The privileging of light skin worked hand in hand with class distinction as relatively elite Belizean Creoles rallied a sense of Belizean nationalism that centered on themselves (Judd 1992). Light-skinned Creole elites' assertion of themselves as a political and cultural force in the colony at the turn of the century reshaped racialization in Belize, further entrenching colorism. Elite Creole men—usually light-skinned as well as economically elite—moved from jobs as foremen of mahogany camps, ship's captains, and retail clerks to the high-status positions of low-level clerks in civil service, as the very small white Belizean population (local whites) receded from political importance (Judd 1992). Less elite Creole people continued to live in rural areas, but those who were lighter-skinned separated themselves out and eschewed interaction, and certainly intermarriage, with darker-skinned individuals. The villages of Burrell Boom and Crooked Tree in particular became known for their light-skinned populations. But privileging light skin and rejecting African ancestry have been widespread throughout rural Creole Belize. Stories about older lighter-skinned individuals refusing contact with darker-skinned ones, like the kiss-kiss story above, abound. Similarly stories of light-skinned individuals disliking dark-skinned people are commonplace. One ancestor of the Rhaburn family is said to have exclaimed that he would never "lie with anyone from Crooked Tree." He was "clear" skinned and did not want to sully his line with darker "blood." So, while Crooked Tree is known for its light-skinned people, it has also long not been "white." Many times I heard the story that the "first black man to come to Crooked Tree" was the renowned Ebeneezer Adolphus, an influential teacher with the Baptist church, who was stationed in Crooked Tree (and Lemonal and May Pen) between 1890 and 1925. Although he was beloved and he married into the community, it is revealing that he is always identified by his dark skin color. The colorism, or privileging of lighter skin, sharper noses, and straighter hair, that characterizes the Caribbean was well entrenched by the turn of the century. At the same time, it is the Belizean Creole, with their mixed African and European (and whatever else might be thrown in) ancestry that assembled with the logwood-rich swampy lowlands of this part of Belize. This tension among Belizean Creoles, between a white bias and Eurocentric, perhaps urban, orientation, on the one-hand, and a sense of pride, comfort, and know-how in the bushy rural areas that were home to free black and brown people for centuries persists to this day; I address this point in more detail in chapter 3.

Living Otherwise at the End of the Nineteenth Century

On the one hand, it could be argued that Belizean Creoles were simply accommodating colonial expectations for them at the turn of the twentieth century: living along distant waterways, cutting wood sometimes, and not developing intensive agriculture. Instead, I suggest that the individuals, families, and communities who made their homes here, did so in order to live as they wanted. That in this way they sustained an orientation that had been put into place a century earlier. The Baymen were known for their independence: "The loggers were independent and doggedly stubborn, rugged individuals who chose to work knee-deep in a malarial swamp for their daily bread, rubbed elbows with alligators, and faced the loss of their homes with every new season, either by flood or forcible eviction" (Joseph 1989, 15, cited in Judd 1992, 49). Rural Creole people today still are often skeptical of authority and government, and one clear indication of this is how land has been used and understood since the settlement's earliest days.

In the earliest days, British woodcutters were legally forbidden from settling and planting in their woodcutting camps. Later land and labor laws continued to discourage agricultural development. Yet these discouragements from governing authorities had little influence on day-to-day life in the lower Belize River Valley and its waterways east and north. From the very earliest moments, logwood cutters made plantashe at their encampments, growing plantains and a host of fruits and vegetables to sustain themselves, while fishing and hunting the plentiful fish and game and wild cattle and pigs in the area. Farming was a widespread avocation for rural Creole people in the nineteenth century. By the late 1800s, the occupation listed by men living in Lemonal, Crooked Tree, Bermudian Landing, May Pen, and the like was often "Planter" or "Cattleman." In the 1870s Henry Fowler, in an account of his travels throughout the colony, noted "small patches of corn, rice, plantains, coconuts and fruits dotted along the river banks . . . which produce sufficient for individual consumption, and any surplus finds its way to Belize. . . . It is estimated that there are 5000 or 6000 settlers in the Colony, yet I cannot ascertain that there are more than about 250 who have any title to their holdings *beyond a squatting possession or mere tenancy at will*" (Fowler 1879, 48–49; emphasis added). This precisely describes the landholding in the villages that interest me; or at least describes landholding from the perspective of governmental authority. For the people living in these places, their houses are located on lands that have been used and occupied by their families for centuries; and they feel profoundly that these lands are theirs. Indeed, the government began surveying in the late twentieth century to formally allocate house lots and lands in the lower Belize River Valley and Crooked Tree areas. Many individuals who make plantashe on larger parcels also used the lands without formal governmental approval for years. Some parts of both Crooked Tree and Lemonal have been claimed by long lost descendants of particular Mahogany works owners at various points in recent history but those claims have been unsuccessful; villagers who have historical connections to their lands have been able to retain those lands, and in recent years have started to formalize those claims (though this has not been a speedy process).

This long-term squatting, along with cobbling together a variety of ways of making money and sustaining oneself as well as the lack of interest in repetitive labor for someone other than oneself are hallmarks of the independent-mindedness of Belizeans living in these places. These modes of living do not conform to capitalist logics of developing the colony, generating agricultural for export, pushing to reap more agricultural product to create capital for reinvestment, or ensuring an available labor pool. Rural Belizean Creole people have always wanted the freedom not to be hurried or rushed, to socialize and be socially connected, and this has trumped conforming to the interests of British colonial authority and the political economy of capitalism.

Into the Twenty-First Century: Conservation, Tourism, and Migration

Although much has changed over the past one hundred years in Belize, and in Belize's relationship to the world—the patterns that had developed have persisted. Becoming Creole continues to mean being entangled with the more than human that assembles in this part of the world, while negotiating a color-class hierarchy in which a bias toward whiteness exists in tension with a pride in connection to this place. This connection is marked by lower class status, being African-descended, and crafting a way of living in the world that is not primarily determined or organized by the logic of capitalism.

Politically, the twentieth century saw the gradual devolution of rule from Britain to Belize, culminating in independence in 1981. Economically, Belize experienced increasing labor unrest that intensified both with the global economic depression of the 1920s and 1930s, and after the return of Belizean soldiers who fought for England in World War I, experienced racism in that context and came home to difficult economic conditions. A massive hurricane in 1931 further troubled the country, and served as a reminder of the power of the nonhuman. The country also benefited from global economic growth in the latter half of the twentieth century.

In Lemonal, May Pen, Crooked Tree, Backlanding, and the other settlements along the lagoons and waterways that interest me, rural Creole people continued their "kech and kill" ways of living, but also developed a reputation for being good farmers, despite colonial officials' contentions about the nature or essence of rural Creole people. Residents of Crooked Tree petitioned for roads to be built into the village so they could more easily transport agricultural products to Belize City, using the argument that they were one of the most successful agricultural villages in the country. Indeed, Crooked Tree became famous for its annual agricultural show shortly after a causeway that allowed access to the village was built across the lagoon in 1983.

Rural Creole people engaged in a variety of new enterprises as their socionatural entanglements shifted. Chicle, logwood, mahogany, and pine continued to be sold, but new items came into market trade. Beginning in the 1940s hunters caught crocodiles and sold their skins until the practice was outlawed due to rapidly declining numbers of the animal. In the 1960s jaguar skins became a valuable commodity, which also became illegal by the 1970s because of concern for the demise of the species. Some rural Creole people made coal beds, producing charcoal from the oak trees that studded the pine-savannahs, and many processed cashew seeds from the trees, which grow well in the sandy soils there. By 1990, the Belize Estate Company no longer cut wood, and there were few woodcutting jobs in the area, but during the fifty preceding years, almost every man in this part of Belize worked for at least sometime in the timber industry; others might have worked in the growing sugar industry at the cane-processing plant at Tower Hill, Orange Walk.

The 1960s might be most powerfully marked by the arrival of Hurricane Hattie in 1961 and the sheer destruction this storm wrought when it leveled not only houses but also plantashes throughout rural Belize. This decade also introduced to Belize a new industry that has been growing since then: tourism. The very earliest tourists to Belize were sport hunters and fishermen. Fishermen came to what was called Keller's and is now the Belize River Lodge. This fishing resort was built by Vic Barothy, an American fly and sport fisherman, and its main location was just nine miles up the Belize River. Barothy also built accompanying lodges on Turneffe Caye off the Belize coast and along Spanish Creek between Lemonal and Crooked Tree at a place called Wade Bank. The latter two are no longer in operation, but the Belize River Lodge persists as a successful tourism enterprise. A number of rural Creole men and women have worked for these lodges as guides, cooks, groundskeepers, and maids. Just north of Crooked Tree, near Revenge, a hunting camp that specialized in jaguar hunts ran for about a decade and employed rural Creole men as both guides and camp caretakers. But in the 1980s, conservation—of birds initially, and then ecosystems and biodiversity more generally—spurred on the nature tourism industry. Belize today markets itself as a tourism mecca for nature lovers. Crooked Tree was one of the first places in Belize to attract the interest of both conservationists and tourism entrepreneurs.

Other changes that have come about in the latter half of the twentieth century include roads and bridges, electricity and water, immigration to the United States, and increasing gang- and drug-trade-related violence in Belize City. Both the causeway to Crooked Tree and the bridge across Spanish Creek to Lemonal were completed in the early 1980s, and bus service runs to almost all the rural communities in the area on mostly paved roads. Electricity came to Crooked Tree in the mid-1990s (though a large generator had supplied nighttime electricity for several years before) and several years later to Lemonal. Water was piped beyond Bermudian Landing to Lemonal in 2015; Crooked Tree was connected to the national water system several years before. With this infrastructure (roads, bus service, electricity, and water), more people are choosing to live in the villages and commute to jobs in urban centers (Ladyville, Belize City,

Orange Walk). In addition, an increase in violent crime in the cities (as a result of the drug trade and the gang violence it generates) has also encouraged movement back to villages from urban areas. Crooked Tree has grown significantly in population and also looks dramatically different from the way it did when I first visited in 1990. Although it could never have been characterized as impoverished, today Crooked Tree feels prosperous in places and is home to many large, beautifully kept cement block houses. At the same time, many villagers live simply on relatively small incomes. By contrast, smaller villages like Lemonal have only witnessed modest change, more cement block houses, more cleared areas. Migration is partially responsible for the prosperity of villages in this part of Belize. Today, roughly half of Creole people born in Belize now live in the United States—Chicago, Los Angeles, and New York are the more likely places to find Belizeans who hail from these waterways, but rural Belizeans from the lower Belize River Valley live all across the United States as well. Many of these migrants live transnationally and build, or contribute to construction, in the villages where they were born.

All these developments have continued to re-entangle Belizeans with the more than human. The elements that are brought into these socionatural assemblages have shifted, and these assemblages now include global processes like tourism, conservation, and migration, rather than simply the vicissitudes of the global timber trade.

Belizean Creole people with ranges of skin color, hair type, and facial features that fall along a spectrum between black and white continue to assemble and gather with the more than human of these places. The racialization of places that began in colonial days continues to associate Belizean Creole with knowing and living well in these landscapes. Belizean Creole people take pride in living here, and their brownness, blackness, and livity come together as a source of strength in the face of dominant global discourses and institutions that dehumanize blackness. These racial assemblages will come to incorporate, in interesting ways, tourism, conservation, and migration, as I show in later chapters.

Finally, even though it might seem that tourism, conservation, and migration bring Belizeans into the neoliberal logic of late capitalism and thereby reorient the people living here, I contend that persisting in being Belizean Creole precisely means reproducing the ways of being their ancestor's enacted. Rural Belizeans are part of socionatural entanglements that include orientations toward sociality and freedom. The words of a rural Creole woman now living in Houston who visited us in June 2016 in Central Texas are instructive. She extolled the "freedom of village life, that is what makes it so sweet." By this she meant the freedom to do what one wants how and when one wants, not to commute to a daily wage labor job, and to be outside and not "locked up inna house," as Belizeans often characterize life in the United States.

CHAPTER 3

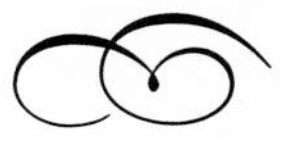

Bush

RACING THE MORE THAN HUMAN

In the late 1990s, my husband and I lived in a nicely kept two-bedroom cement block house in Ladyville, the ninth largest population center in Belize, suburb of Belize City, and home to the country's main international airport, yet a place that is still called a "village." Ladyville is located on the coast, in the midst of mangrove swamps cleared and filled to create land solid enough for development. The land is swampy, and our yard was full of lush green grass that grew quickly and that my husband always kept trimmed short. It also contained a tree that shed leaves often, and after windy storms, a flood of small branches and other detritus were inevitably strewn across the yard. We frequently had family and friends over to visit and on the mornings after these impromptu gatherings, the yard would be scattered with a few stray candy wrappers and empty cigarette boxes and butts, along with the glass beer, rum, and soda bottles awaiting collection and return to the bottling company. One morning, when the tree had shed and after family members had visited, my husband and I we were sitting outside, surveying our yard. Noticing all the paper, plastic, and glass, I said: "Wow we really need to clean up the yard." He replied, "Yah, di yard only bushy; I wa borrow a rake and clean up these leaves and sticks." I had not even seen the leaves and twigs, my eyes focused only on the bottles and candy wrappers. I said, "I meant the paper and glass," and he said, "Oh yeah," as if seeing those for the first time. In rural Belize, unruly yards beckon scorn. In 2013, an SUV full of my Belizean relatives and me, representatives from three or four different households and yards, drove through Hattieville, a village outside Belize City on the Western highway. As we traveled, we were each noticing and commenting on different plants in the yards we drove past. When we passed overgrown, bushy yards, everyone burst out into gently scornful laughter, and then teased each other about how bad each others' yards looked, how they were "only bushy."

As Caribbean cultural theorist and literary figure Eduard Glissant suggests, landscape itself is a character in the process of making history (DeLoughrey 2011). In this chapter, I examine the character of "the bush," a Belizean Kriol concept that describes the tangled subtropical forest and scrubland in which rural Creole people have made their homes. The concept has a host of meanings and associations that make it particularly interesting, and also yield insight into how socionatural assemblages always articulate with a variety of discourses of value and hierarchy. The bush assembles with blackness, and is further associated with being backward in some contexts, and with competence, strength, and freedom in others.[1] These associations are necessarily never stable but jostle with and against one another in different times and places. The meanings associated with the bush emerge out of internal conversations within rural Creole communities, within the larger nation of Belize, and they articulate with global orders of race and color and transnational discourses of tourism, conservation, and national development.[2]

The term "bush," describing wildlands, has a long history and is widely used in white British colonial settler societies—from Africa to Australia and the Caribbean (Anderson and Grove 1987; Dominy 2001; Dunlap 1999; Ross 2008).[3] While the use of the term "bush" is not limited to tropical places (e.g., Alaskan bush), it seems to be more fully developed in descriptions of the tropics.

The word shows up in a variety of texts that reveal the meanings associated with it. For example, Frantz Fanon in the opening to *Black Skin, White Masks* writes: "The more the colonized has assimilated the cultural values of the metropolis, the more he will have escaped the bush. The more he rejects his blackness and the bush, the whiter he will become" (Fanon 2008 [1952], 3). Similarly, in the highly acclaimed novel, *Americanah*, about young Nigerians who migrate to the United States and the United Kingdom and then return to Nigeria, Chimamand Ngozi Adichie's characters use the term "bush boy" to describe a young man from a village far from any cities in Nigeria, who has the desire to be as urban and sophisticated as his friends at an urban Nigerian high school (Adichie 2013, 305). In both uses, the word describes untamed nature and a distance from the metropole, and Fanon expresses its connection to blackness.[4] Closer to Belize, Aisha Beliso-De Jesús describes the association of blackness and Africanness with remote, wild "bushy" areas in Cuba, noting that these locations serve as the symbolic home of the Afro-Cuban blackened set of spiritual practices called Santeria (Beliso-De Jesús 2015).

Eurocentric racial formations and the antiblack racist ideology they generate are implicitly woven into the meaning of the bush. In this logic, the civilized, sophisticated, knowledgeable pinnacle of whiteness is counterposed to the savage, animalistic, antisocial, uneducated nadir of blackness (Anderson 2003; Jahoda 1999; Khan 1997; Mikulak 2011). This binary is often mapped onto space and, as numerous critical race theorists assert, race and space are always mutually constituting (Gilmore 2002; Lipsitz 2007; McKittrick 2006; Mills 1997). Charles Mills notes that in the racial contracts of Eurocentric states, and in the systems of being and knowledge that enable those contracts, space is racialized at the level of country and continent, at the local level (city and town), and at the micro level of the body itself (Mills 1997, 43). He also shows how the idea of the savage, the creature of the forest (the word "savage" is rooted in "silva," forest) is closely implicated in these processes. The white supremacist racialization of blackness entails locating this embodiment closer to "nature" and farther from the center of "civilization" that the white racial frame constructs as the fulcrum of what it is to be human (Kim 2015).

In Belize, where groups of enslaved men of African descent cut mahogany in forested areas away from Belize City and where free people of color could find places to settle "off the main" in the nineteenth century, these bushy places became associated with African-descended peoples, reproducing the closer-to-nature and less-than-human constructions that the logic of white supremacy attributes to dark-skinned people. The late twentieth century brought new twists to these assemblages of race and nature with the rapid growth of ecotourism and nature conservation. People living in rural Belize have become subject to the ecotourist gaze and the logic of biodiversity conservation projects that in new ways have assembled together bodies, racial codes, animals, and plants. The more than human of these tropical landscapes is often described as empty of people in ecotourism advertisements and conservation literature. The only ones who might be present in this discourse are indigenous peoples, who are cast as "ecologically noble savages" (Redford 1991), but blackness is not typically associated with tropical forest conservation. Conversely, these bushy places are now sites for the attention and visitation of whiteness—the state of being an ecotourist or a biodiversity conservationist assembles strongly with whiteness (Bandy 1996; Braun 2003; Fletcher 2014; Mowforth and Munt 2003; Urry 1990; Werry 2008).

While the white racial frame plays a role in how rural Belizeans understand themselves and the bush and the rural-urban divide, that is certainly not the only organizing frame. As Sylvia Wynter states (quoted in McKittrick 2006, 123), there is "always something else besides the dominant cultural logic going on . . . not simply a sociodemographic location but the site of a form of life and of possible critical intervention." In theorizing plantation society in the Caribbean, Wynter has pointed to the "plot," where enslaved Afro Caribbean people grew their own foods apart from the plantation, as one such site for these critical interventions (Wynter 1971). Michaeline Crichlow and Patricia Northover similarly suggest that "fleeing the plantation" and re-rooting in place, or making place, are practices of freedom (2009a, 19). Here, then, blackness and brownness can be assembled differently and can proliferate beyond the confines of the white racial frame. Rural Creole people have assembled knowledge, skill, and well-being with their embodiment as non-white, and the trees, animals, plants, waterways, winds, and other more than human among which they live. Their capacity to thrive in these areas assembles blackness and brownness with freedom and livity, or the capacity to create their own livelihoods at least partially on their own terms.

The Bush in Belize

In Belize, the words "bush" and "bushy" and their various connotations reveal a great deal about the ambivalent and complex relationship that is rural Creole socionature. The bush is the wild, sometimes densely vegetated, sometimes not, land that expands out beyond the inhabited areas of villages. The term "bush" is relative—compared to Belize City, all villages in Belize are bush. Yet an uninhabited dense stand of cohune palm forest, for example, is bush in relation to the village center, where houses are concentrated. Richard Wilk has described the meanings of the term: "In the colonial geography of Belize, the bush was at one end of a continuum, and the city at the other. The bush was wild and primitive, while the city was considered civilized by virtue of its closeness to European ordered space, and the number of Europeans and European institutions based there . . . this spatial order was mirrored in a time scale which placed the bush in a distant . . . past and the European cities in the present" (Wilk 2005, 5; see also Wilk 1989).

As much as the term is used to describe landscapes, it is also used to refer to particular people, or ways of doing things, talking, or thinking. When the term is used to describe individuals, it is often used in jest, in joking insults that typically re-enforce particular social hierarchies. However, the term can also be positively claimed. The exclamation: "I live da bush!" can be a point of pride, calling dominant hierarchies into question, celebrating the positive values of competence and freedom that mark rural Creole culture. All rural Creole Belizeans are bushy by virtue of being rural. Those who come from smaller and more remote villages are more bushy than those who come from larger villages more "on the main," and Belizeans living in Belize in general are seen as bushy compared to people living "da states." While the term has long been used in this place, it also always reflects a global gaze, whether from the British colonial officer, the Houston-based ecotourist, the migrant living in the Bronx, or the middle-class but worldly Belize City urbanite. In the remainder of this chapter I explore how this term is used to describe individuals, practices, and landscapes in turn.

On Being Bushy

The term "bushy," used either as an adjective or a noun, is a commonly used epithet and descriptor among rural Belizean Creoles.[5] It is often used as a playful insult, but it is also sometimes embraced in a playfully ashamed or defiantly proud way. Most commonly, bushy-ness has relatively negative connotations. Being bushy means being "backwaad" (backward) and far from the metropolitan center of power. For example, a young middle-class Belizean Creole relative, Sam, who lives in a relatively wealthy urban area in Belize, was visiting our house in 2015, and did the dishes in ways that we normally do not do dishes. My husband, half joking, but also wanting to make clear how Sam should do dishes next time, looked at him, made a short comment about what Sam was doing, and then said, "You mus coh fa *bush*" (You must be from the *bush*). Sam and all within earshot chuckled. Sam, who does not spend much time in rural villages, shook his head in mock wonder at his uncle who would make such an "idle" (useless and trouble-causing) comment.

Overlain with these meanings are racial overtones. This emerges in a prominent way in the epithet "bushy-headed" that I have heard used both playfully and scornfully depending on the context, but which always means having unkempt, "natty" (kinky) hair. Common derogatory phrases and commands in which kinky hair is equated with bush and shaded as African and black include: "Yu head *only* bushy" (Your hair is so bushy); "Yu, bush-head, coh ya" (You, bush head, come here); "Yu head bushy *bad*, gyahl" (Your hair is really bushy, girl); "Bwai, go fix your bushy head" (Boy, go comb your bushy hair). Both hair and tropical vegetation can quickly become unruly—keeping bushy-ness at bay is a constant struggle.

The words "bush" and "bushy" are associated with a number of related words. One of these is "wile," which in Belize Kriol means being shy of, or timid around, anything new or unknown, wild the way a deer is wild. "Ih wile like ih coh fa bush" (He's wild like he comes from the bush) is the kind of statement that illustrates this connection. A phrase that often accompanies being bushy is "dumb to di world," which means naive, unsophisticated, and unable to navigate city life. Another common word that can sometimes be used interchangeably with bushy is "backwaad," which connotes the opposite of modern and civilized, words that are sometimes used, but much less often. These connotations combine together so that the cosmopolitan urban person, who sports neat straight hair, nice clothes, and knowledge of the world stands in diametric opposition to the bushy, who is "wile," "dumb to di world," and "backwaad."

The racial overtones of bushy are also highlighted by how easily bushy can be interchanged with the Belize Kriol phrase, "new neega." This is a less used, but punchy and chuckle-drawing epithet that translates as "new nigger" and describes someone so fresh from Africa, referring back to slavery days, that they have no knowledge about the world. I have often heard apocryphal stories told for amusement about Belizeans who have migrated to New York City or Chicago in November and jokes being made about how they were so new neega that they think all the trees have died, because they had no knowledge about leaves dying and falling all at once in autumn in temperate climes. Or in another story, the Belizean does not know what that white dust falling out of the sky in winter is. A Creole person might laugh at themselves as "new neega-ish" if they are still wearing the tags on their clothes if they put on a newly purchased jacket—with the implication that a desire to boast new clothes reveals that the person does not know the social norms in cosmopolitan places of not wearing clothing tags. All these stories emphasize a lack of knowledge and familiarity with the world outside of rural Belize, and imply that people who are bushy are backwaad and not modern and sophisticated.

The connotations of bush can also bring together blackness, strength, competence, and the more than human. For example, although they may be backwaad and wile, people who come from the bush are also independent, tough, and able to survive in any circumstance. They are "haad," not "soff," in Belize Kriol. The contrast between haad and soff is frequently referred to by rural Belizeans. Being haad is good and being soff is most often not. If you are haad, you can fend for yourself, and you will not get stepped on by other people. If you are soff, you are dependent on others, easily hurt, and weak. People who come from towns and cities, and in particular the middle classes, are soff in the eyes of rural Belizeans, as are those identified as "bakra," or white.[6] This hardness and softness itself is racialized and linked in a very material way to skin. White skin is soff, easily burned, and easily hurt by the physical world, like the people who bear it. By contrast, black or darker skin codes the capacity to handle physical hardship with grace and competence.[7] White people from the United States and Europe may be powerful, rich, and influential, but if disaster strikes they will not survive, as one older Creole man explained to me in the 1990s. Being from the bush, and thus implicitly more closely allied with blackness, then, is a kind of badge of toughness, of overall competence in life. Therefore, at the same time that one person might use the epithet "bushy" to jokingly denigrate somebody, another person might say with defiant pride "I come from the *bush*," assembling positive associations not only with particular landscapes but also with brownness and blackness.[8] Rural Belizean Creoles are proud of their competence in the bush, of having the physical and mental wherewithal to handle "life da bush."[9]

The positive valence of bushy-ness and being haad is exemplified in the pride that rural Creole women take in their ability to process cashew seed, an arduous and potentially painful task. The production of cashew nuts is difficult and labor intensive.[10] The husks of cashew seeds contain oils that burn and damage human skin (often discoloring the skin). Women char the husks in order to be able to easily peel them off. While some women use gloves to do this, most use their bare fingers, which then peel and become stained with various shades of red and brown. Women will point to their stained fingers as a testament to their strength and hard work, to their competence in the natural world, and to their hardness, proud of their identity as rural Creole women. Rural Belizeans' relationship to the more than human, their bushy-ness, might mark them as lesser than urban and middle-class Belizeans, and might be racially inflected with a blackness that they do not always embrace, but it also bonds them together in a shared pride in toughness, competency, and pride in place.

Bushy Foods

Food is a rich arena in which social hierarchies are enacted and reproduced (Mintz 1996), and ideas of the bush inhere in food-related hierarchies in Belize. Richard Wilk has plumbed the cultural meanings of food in Belize in numerous publications (Wilk 1999, 2006). One story he shares is particularly illustrative of how ideas of backwardness and cultural identity are tied to the bush—or at least to foods that are considered to be bushy. In 1985, Queen Elizabeth II of England visited Belize, where a state banquet was held in her honor and the best cooks prepared local delicacies for her. One of the best-tasting meats that comes from the Belizean forest is gibnut (tepezquintle), a close relative of the guinea pig. Although this has long been a popular meat in the rural diet, it has only become desired among the urban middle classes in the past thirty years or so.[11] The gibnut was chosen as a delicacy that well represented Belize's newly emerging local cuisine and was proudly served to the Queen. The headlines in the papers in England the next day boldly proclaimed "Queen fed Royal Rat." Belizeans were put in their place for their bushy-ness and their backwaad food choice. These headlines provoked outrage in Belize and had the effect of creating a new interest in and acceptance of bush foods as the most authentic Belizean food, even among urbanized middle-class Belizeans. Although I was not living in Belize when the Queen visited, the story of her visit, and her eating gibnut, was, and still is, often told. Each time I heard the story, the Creole storyteller expressed pride in the fact that the Queen ate gibnut and amused dismay at how the event was covered in the British news. By the mid-1990s, a return rural Creole migrant born and raised in Crooked Tree but living in Los Angeles at the time told me of her dream of starting a restaurant for both tourists and villagers that would serve "local food like gibnut and hamadilly (armadillo)." She wanted to exalt "life da bush" for the tourists and allow them, as she put it to "enjoy the raw life of Belize itself."

However, the tension between celebrating village life and being proud of the bushy nature of Belize, and the desire of Belizeans to be "civilized," as I have heard people put it, persists.[12] Bushy practices are tinged with a lower status, even if they might be "authentically Belizean." This point is well illustrated by a news feature from 2003 that I relate below, as well as by the social media conversations I discuss in chapter 7. The news feature I share was done on the first Tilapia Festival held in Crooked Tree in late March 2003. The reporter, a young woman with the news station Channel 5 Great Belize Productions in Belize City visited the village on the day of the festival and spoke to a variety of villagers about the recent popularity of tilapia as a food fish. The feature ends with this exchange:

> We did manage to find one man whose tastes ran to something more exotic . . . a plate full of iguana eggs.
>
> NEWS REPORTER JACQUELINE WOODS: "How it taste?"
>
> KARL VERNON, RESIDENT, LORD'S BANK VILLAGE: "It's most marvelous. Beautiful I have to say."
>
> JACQUELINE WOODS: "How it taste, like boil egg?"
>
> KARL VERNON: "Yes. Want to try one? It's very delicious. Tek anyone you like."
>
> JACQUELINE WOODS: "Trust me guys, this is the first time I'm going to try this."
>
> (Jackie tries to bite egg) "I can't bite into it though? It kinda hard."
>
> KARL VERNON: "Bite the pint (end) man, bite the pint and mek a hole in it. Now squeeze it out" [Bite the end, and make a hole in it, now squeeze it out].
>
> JACQUELINE WOODS: "It noh taste too bad you noh, but just the thought that I'm eating an iguana egg, I think that dah weh got me kinda nervous" [This is what has me kind of nervous].
>
> (Great Belize Productions 2003)[13]

There are two key things to note about this exchange: first, the interest in the "exotic" (although iguana eggs are not such unusual fare in *rural* Belize); and second, how the Belize City-based reporter positioned herself as having difficulty eating the egg (they are not difficult to eat), having never tried it before and being nervous about trying it. Through her body language and comments, she marks very clearly that this is *not* something she is used to. In this case, gender norms also shape the reporter's encounter with iguana eggs. Her performance of gender here associates elite status and urbanity with a delicate femininity and these are shaded with whiteness and holding herself distant from the bush (see Ulysse 1999).

The multiple and contrasting meanings of the bush are well illustrated by these two vignettes about food practices. On the one hand, rural Creole people are proud of how their particular bushy foods make them distinct, but at the same time these foods, and their bushy-ness continue to mark them as backwaad and this itself is tinged with a negative association of blackness.

The Bush as Landscape

The bush describes a landscape and, in terms of landscapes, the state of being bushy is relative, as noted above. All of Belize is bushy compared to States. And indeed, the Belize Kriol term "States" really only includes urban, inhabited places with bright lights and places to shop. I was driving between cities in Central Texas at Christmas in 2015 and one of my passengers was a Belize City born-and-raised man who had been living in Chicago for the past twenty years. When we passed through a long stretch of rural Texas plains, with no shopping centers, city lights, or hustle and bustle, and hardly a building, he remarked on this landscape, saying: "That's not the real America. It's not like you are in America."

The whole country of Belize's location as bush is illustrated by an early 2016 video-chat conversation in my home in Texas between a middle-aged rural Belizean-born-and-raised-man living in the United States and a young man living in one of the more elite areas of Ladyville, a suburb of Belize and home to Belize's middle class. The U.S.-based Belizean wanted to see a dog that was being kept outdoors behind a house in Ladyville, but the video conversation was being held at night, and it was dark out. The young man in Ladyville said, "Wait, I need to find a flashlight." The middle-aged U.S.-living rural Belizean responded, "Sounds like you de da (are in) bush"—the crowd gathered around the video chat in the United States all chuckled. The young man responded, "All of Belize da (is) bush," and everyone laughed harder, agreeing.

While the laughter in this conversation reveals some of the implicit sense of "backwaadness" encoded in the idea of the bush, some comments about how bushy a place is have more pointed and explicit moral implications. Take, for example, the villages of Crooked Tree, Lemonal, and the central-lower Belize River Valley communities of Bermudian Landing, Double-Head Cabbage, Willows Bank, and Flowers Bank, among others. Crooked Tree is a large village located on sandy soil, with plenty of naturally open grassy space, and wide-open pine forests. Residents pride themselves on the open, clear views, or in other words, the lack of bush. Many people in Crooked Tree are also very light-skinned, and the village has a reputation for its beautiful light-skinned girls.[14] By contrast, the villages of Lemonal and the lower Belize River Valley are smaller, located on denser black soil that gives rise to thick subtropical forest, and are less open. Few of these villages are renowned for their predominantly light-skinned residents, although all of them are home to people with a wide range of skin colors.

When I was living in Crooked Tree, a number of people with whom I spoke told me that they did not particularly like other villages, and that I would not like them either because they were "very bushy and full of mosquitoes," as one person told me. They contrasted bushy villages with Crooked Tree, which was instead "a nice and peaceful community and beautiful too." The implication, of course, was that the more bushy villages were not as nice or peaceful. People in Crooked Tree warned me of the danger of neighboring bushy villages that had high bush and muddy, rather than sandy, roads. These places are "behind time," so far "back-a-bush" to be beyond redemption. The people in these places speak the "wussus a Kriol" (the worst Kriol), or the most "raw," the version of the lingua franca that is least like English, and contains more words and grammatical structures from West African origins. These comments were always made lightly, at least partly in jest, but they nonetheless illustrate rural Creole place ranking on the basis of how bushy and backwaad they are.

If being bushy connotes being backwaad and not cosmopolitan, then cleared bush is an indicator of being modern and civilized.[15] Belizeanist scholar Laurie Medina shared a commentary she heard a rural development officer make during a radio interview in the 1990s. Speaking to a national audience and lamenting the state of things in Belize, the officer proclaimed: "When I travel along the highway from the villages to Belize City, I see bush!" The implication is that this bush is unproductive, wasted land, evidence of Belize's backwardness (Medina, personal communication).

Villages in rural Belize have been becoming less bushy over the years. They are "building up" and most have more areas with houses and more cleared stretches of land than they did twenty years ago (see figure 3.1 for what clearing high bush looks like). The magnitude of these changes was made clear to me in Crooked Tree in the mid-1990s, when a man born and raised there, but who had been living in the United States since 1955, returned for a visit. He had been gone forty years. He came to the house where I was staying and we had a long chat. He said he was shocked and overwhelmed; he did not recognize the place at all as the village where he had grown up. As he and another older man who had never migrated out began to describe how the village looked when he had left, I came to fully appreciate his shock and to better understand how bushyness is associated with a long gone past. In the mid-1950s, all the houses in the part of the village where I lived in the mid-1990s were thatch and pimenta—made from the slender trunks of a small palm tree with woven leaves for a roof. There were no roads, only long narrow pathways, "picados," or small paths chopped through the tangled tropical foliage: "The whole place was bush, dense bush," the older man said. In 1995, when the return migrant visited, all the houses

Figure 3.1 Clearing bush in Lemonal to make a pasture, Lemonal, 2007. Photo by Elrick Bonner.

were made of wood or cement and only a few pimenta houses remained. Yards were large and "cleaned down," and roads wide, flanked by well-maintained stands of cashew trees. By 2013, almost twenty years after this man's visit, the village was more like a town, cement houses were the norm, and some newer ones looked more like mansions than houses. On what had been the outskirts of the village, where narrow footpaths were common, there were now many intersecting, wide roads, yards chopped low, and houses—either finished or under construction. Now the bushy outskirts—typically pine savannah in Crooked Tree—were very far from the village center. There was even a brand-new causeway across Western Lagoon. For older people in Crooked Tree, the way the village has been changing, with larger expanses of well-maintained grassy yards and open roads, is material evidence of how Crooked Tree has become less backwaad.

Similar stories arise in other rural Creole villages. My father-in-law described Lemonal when he was a boy in the 1940s as a scattering of houses in the high bush (dense and tall subtropical forest). "Lone lee picado roads" (only little narrow bush-encroached pathways) were available to walk or ride a horse down, connecting houses with Spanish Creek, when travel by boat was most common.[16] There were no bridges across waterways, people used boats to ferry themselves across. Today, wide roads, some even paved, and a large and beautifully engineered steel and cement bridge crosses the deep and dark waters of Spanish Creek. A frequently chopped-down cricket field and large cement or wood houses make up the landscape. Such shifts represent the increasing "progress" of the village from its bushy and backwaad past to its cleaned-down, modern future.

Just as being from the bush has both negative and positive associations for individuals, bushy landscapes, while most often negatively valued, also sometimes have positive valuations. Often when people would make comments about how bushy and unpleasant a village was, favorable comments would slip in. They would say of the most bushy of villages that, no matter where you find yourself at mealtime, you will be fed. People in villages back-a-bush are very generous, even if (or perhaps because) they live in these places. The implicit suggestion here was that as places became less bushy, people were less generous. Occasionally, I heard rural Creole people from villages that are seen as most bushy make critical comments about the haughtiness and arrogance of villagers from more open places, implicitly opposing this to the greater humility and groundedness of people from villages that are more bushy.

Freedom

A commonly expressed positive association with the bush is that of "freedom," a word I heard used many times to describe what people most liked about living in these villages. Rural Creole Belizeans highly value their freedom.[17] The salience of ideas of freedom for descendants of enslaved Africans and English slave owners may be that much more powerful. Ideas of freedom are linked to the importance of *land* to Afro-Caribbean peoples.[18] In Belize as in other parts of the Afro-Caribbean, emancipation from slavery was immediately followed by the imposition of restrictive property laws designed specifically to prevent newly freed slaves from acquiring land. This prevented the rise of an independent peasantry and instead ensured the ruling class a large pool of low-wage workers. For those individuals who managed to secure a piece of land, that land was a powerful tool (both materially and symbolically) in the struggle for freedom.[19] Scholars of the Caribbean argue that this history has led to land symbolizing freedom for Afro-Caribbean peoples and a strong Afro-Caribbean attachment to the land, manifest most clearly in the institution of family land, a plot of land passed down through generations and shared by family members (Besson 1987, 1993; Lowenthal 1961).

Most rural Creole Belizeans certainly feel a deep attachment to the land, or more accurately, to the *place* of the villages where they were raised, even if they happen to live elsewhere as adults. These villages sit in the midst of biodiverse subtropical forests, scrublands, and wetlands that provide a wide variety of foods to eat, material goods to use—a means toward living independently and sustaining themselves. Furthermore, the distance of these bushy villages from the center of government and from government reach, provide a degree of freedom from government meddling. Rural Creole people resent any government-imposed limitations on their hunting and fishing, or any requirements for them to pay taxes on land that has been dwelled on by their ancestors for hundreds of years, for example. And many enjoy living in these communities because they can organize their time as they wish while they work to sustain themselves from the more-than-human plenitude around them.[20]

People from these places frequently draw a contrast between the freedom they have in their villages with the lack of freedom that characterizes life in Belize City or the United States. Indeed, people often voice the criticism that living in the United States means that one is "lock op inna house all day, how could that sweet?" (stuck inside a house all day, how could that ever be pleasant and fun?). Living in Belize, and especially rural Belize, people can move around and be outside.[21] Living in or close to the bush gives people the freedom to hunt, fish, make plantashe, and generally do as they please. Unlike the work associated with slavery or wage labor in the mahogany camps, for example, which has to be done at certain times and in certain ways, and entails repeating the same kind of work over and over; being able to make a "kech and kill" living allows people to work for themselves on their own time—to not have a boss, and also ensures that one will engage in a wide variety of activities to "make do" (Bolles 1996). In the bush, then, or at least in places closer to the bush than Belize City, rural Creole people can to some extent bring into being ways of living *otherwise* that operate on logics different from those of the regimented and time-clock orientation of the global capitalist economy that is always also a part of this place.

Cleaning the Village, Removing Bush

Rural Creole Belizeans work hard to clean down bushy areas in public and community areas throughout their villages. Cleaning down these bushy spots and removing unkempt shrubs, grasses, and small trees creates aesthetically pleasing landscapes and marks rural Belizeans as modern members of a global society.[22] By maintaining these open and cleaned-down spaces, Belizean Creoles assemble themselves with all that is "modahn" and "civilized," the words Belizeans occasionally use to contrast with the much more commonly used "backwaad."

With the onset of ecotourism in the 1990s, villagers in Crooked Tree became increasingly focused on the appearance of the landscapes in which they lived. Long-standing concerns about the village's bushy areas began to be seen from the imagined view of the white tourist from States, and the urban middle class or rich Belizean from Belize City and Belmopan, whose attention has been turned to rural areas with the ecotourism boom. There are places in villages that become bushy quickly: small gaps between yards and along roadsides, or in the occasional yard where no one lives, or where people too old or too marginal to care let bush grow up.[23] This small patchwork of bushy-ness became the focus of much discussion in Crooked Tree village council meetings and around kitchen tables as ecotourism began to blossom.

The community burying ground in the center of the village was one of the first sites to garner concern. In the 1990s, the graves and tombs lay under the branches of a huge, ancient "tubroos," or Guanacaste, tree with a large spreading crown. The ground at the edge of the cemetery and the areas between the tombs and graves would rapidly become overgrown with bush that was only occasionally chopped down. Since this was in the very center of the village, any tourist driving or walking in the village necessarily passed close by. And because this was a common area of the village, no one person was responsible for keeping it chopped low. In spring 1994, Crooked Tree hosted a group of study-abroad students from the United States for a weeklong stay. One of these, a young white woman, was a budding photographer. After having been in the community for a few days, she stopped to take photographs of the burying ground when the bushy areas within and around it were especially high and unkempt. She was excited about her photograph, and she beamed when she showed it to me—the composition was beautiful and to both my and her white U.S. middle-class eyes, the image was aesthetically pleasing. To her eyes, this was yet another way in which the village was charming and beautiful, the wispy inroads of nature into the cemetery reminiscent of her brief incursion into the natural world on her boat trip through the lagoon to watch birds. The scene fit into discourses of fading rurality and quaint backwardness, and the ways in which rural areas can be places out of time—an association that, as Fabian has shown us (albeit in a footnote afterthought), is also racially coded (Fabian 2014 [1983]).

A village council meeting had been scheduled for a few days later, and one of the first things discussed was this young photographer. The village chairman at the time, a charismatic young man, Fred, expressed his concern. Fred had noticed her taking the picture and was dismayed. He lamented to the assembled villagers how the burying ground was so bushy and that a tourist had scandalously taken a photograph of this scene. He said, "Dis da pappy show," meaning that this picture taking was like the village standing up and making a fool of itself to the world, and being too stupid (fool-fool) to know better. For Fred, this was a tourist taking a picture of "the village at its worst." The bushy burying ground cast Crooked Tree as backwaad and out-of-time. Shortly following this meeting, a group of villagers carefully chopped and cleaned the cemetery.

Other bushy areas also occupied discussions at village council meetings. Tangled growth near the community center building particularly incensed villagers. At a village council meeting in 1993, a village leader, Mike, said, "It is a shame to have a total jungle right by the community center. Crooked Tree is a village of educated people. We must stop this way of life. It is not natural for people to live in jungles."[24] This comment was followed by remarks from several other people who spoke about the dirtiness of the lagoon area, which was strewn with fish bones at the time. The village council chairman described how a group from a civic organization in Belize City had wanted to come picnic in

Crooked Tree, but "the ground around the church was so terrible, bushy and all" that they did not come.

The complaint and the comments that followed it depended on the dichotomy between bush and sophistication; the assertion that Crooked Tree is "a village of educated people" implies that educated people do not live in the bush. The leader's call for "stopping this way of life" registered dissatisfaction with a backwaad way of living. Progress entails looking the part, making a place look as if it belongs to the global community of educated people—consisting of fellow Belizeans and ecotourists—who do not live in the jungle. The village-wide concern about how the more than human intervenes in the place of Crooked Tree reflected villagers' desire to make Crooked Tree a place that people from elsewhere in Belize and all over the world visit and admire.

Since the mid-1990s, Crooked Tree's population has grown and the village has become much less bushy, as described above. Fewer public areas have the potential to become bushy. Yet keeping the bush out of yards is always a challenge, and Belizeans always notice how well their compatriots tend their yards.[25]

Cleaning the Yard

Yards are a central cultural institution in the Afro-Caribbean, and their significance extends back to the days of slavery. In the plantation Caribbean, enslaved people used small parcels of land on the edges of plantations to grow food, and in some cases would sell that food in markets. According to Sidney Mintz, the yard can be seen as "an accomplishment of Caribbean peoples during their point of sorest trial" (Mintz 2010, 11), or a testament to the power of the human spirit and the ability to create in times of hardship.[26] Indeed, Jean Besson and others contend that the culture-making that occurred in yards was critical to the emergence of Afro-Caribbean cultures, and remains central to their reproduction (see also Austin 1984; Bush 1988; Chevannes 2001; Walcott 1992). The yard is a site where Afro-Caribbean people engage the more than human in culture-making, affirming themselves, and generating livity. Although slavery in Belize was not plantation-based, the importance of the yard—the parcel of land on which people lived—was equally essential to the development of Belizean Creole culture. Here, rural Belizeans created cosmopolitan spaces in the midst of dense tropical vegetation, assembling blackness and brownness with the more than human in particular ways.

Most people in rural villages have large yards. Traditionally, each yard would be home to multiple buildings, some on stilts, some on the ground. These would include at least one building for sleeping (ideally, if the family was well-to-do enough, on stilts) that had at minimum a "hall," or large open common room for socializing, and a bedroom for sleeping. The yard would also contain a separate kitchen structure with an adjacent outdoor fire hearth, a bathhouse, and an outside toilet. Many yards are also fenced. As cement houses have become more common than wooden ones, and with the advent in the 2000s of piped water and electricity, more yards only have one single cement block house complete with kitchen, bath, and toilet in the middle of a large piece of cleared land. Rural Belizeans also put effort into beautifying their yards with decorative and flowering plants, and proudly keep a variety of fruit trees that bear throughout the year (see also Shillington 2008; WinklerPrins and Sousa 2009).

Nearly every inhabited yard in rural Creole villages is chopped "clean and low." People work hard to keep their yards free of high grass and bush, are admired for the work they put in, and take pride in their clean yards. Because so many Belizeans migrate out of their villages, live at least some of the year away, or are in the often multi-year process of building a new house on a newly cleared piece of land, many yards are not inhabited. These are likely to become overgrown quickly and the migrant or homebuilder must send money to pay to have the yard chopped and mowed, or ask family members or friends living in the village to do the work of clearing the yard.

When my family visited Lemonal in December 2009, an overgrown yard near the center of the community precipitated a great deal of commentary. The house belonged to an elderly woman who had spent the past year with some of her children in the United States, and the yard had not been chopped in a long time, trees were overgrown and shrubs and other vegetation had grown up along the edges. Neighbors, many of whom were close family of the woman, gathered together during the Christmas holiday to cut down trees and clean the yard, restoring it to a proper state, so that it no longer looked "terrible," as I heard many people proclaim its previous condition. Not only were the residents of Lemonal pleased that a yard close to the center of the village was no longer bushy, neighbors were pleased to have longer, clearer views across a cleanly mowed and cleared area.

In 2013 during the rainy summer months, I spent time at our family home in Lemonal that sits on another large yard in the center of the village. A group of friends and family gathered to clean the yard shortly after we arrived in preparation for an upcoming birthday party for a young boy in the family (see figure 3.2). Mowers, machetes, and rakes all were employed—mostly by men, but not exclusively.[27] Within several hours, the bushy parts on the edges were chopped low, the debris raked up and hauled away, and the grass throughout the large yard was mowed low. Five-gallon buckets were hoisted by children to different parts of the yard; and at each place young boys and girls, including my sons, picked up the plastic and paper trash, and the "pint bottles" (soda and beer bottles that would be turned in for a discount on a new tray of glass-bottled soft drinks or beer). The buckets full of paper and plastic were hauled and dumped to the "dirt hole" (garbage hole) behind the outdoor latrine. To everyone's

Figure 3.2 Cleaning the Bonner family yard. Author's sons and sister-in-law in foreground, Elrick Bonner in background, Lemonal, summer 2013. Photo by Melissa Johnson.

delight, at the end of the cleanup, the dirt hole was doused with gas and set on fire. The yard was clean, the trash disposed of, and the soda and beer bottles collected. The family and friends who did the work were proud and pleased with the results. The yard was now an inviting space and throngs of children played cricket on the newly chopped lawn.

Keeping yards and community spaces cleaned down is an important dimension of rural Belizean Creole culture. Subtropical vegetation grows quickly and keeping yards clean and pastures useful for cows and horses entails constant work and a deep awareness of the power of the natural world. The importance of keeping the bush at bay, keeping a place looking good in a global discourse that casts bush as negative, is made that much more pointed by the power of the natural world to always overwhelm. Rural Belizeans focus energy on mitigating the capacities of the more than human to become intertwined with and ultimately overtake human created space.

Conclusion

The bushy socionature through which people become rural Belizean Creole is racialized within the logic of white supremacy that casts these marginal areas as backwaad. Keeping the bush at bay has been a central concern of rural Creole Belizeans as they create the landscapes they call home. Chopping yards and public spaces low produces rural Creole Belizeans who are civilized, modern, urbane, and not "kroffy bushies" from "back-a nowhere." The focus on creating cosmopolitan landscapes has only intensified under the ecotourist gaze that continually casts rural Belizeans (and any peoples living in a nature tourism locale) as backwards and traditional. At the same time, in rural Belize, the bush assembles blackness and brownness with ideas of competency, place-based identity, sociality and freedom. Rural Creole Belizeans generate personhoods *otherwise* through their entanglement with bushy socionature

CHAPTER 4

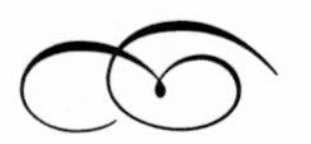

Living in a Powerful World

In April 2010, my husband, Elrick, and his sisters, brothers-in law, and cousin, most of them with extensive experience in the bush, embarked on a fishing trip from Lemonal to New River Lagoon. They traveled in two trucks along a deeply rutted dirt road for several miles through broken ridge (subtropical forest), pine-dotted savannah, and then reedy dried swamplands interrupted by sharp-needled calabash trees to reach the lagoon edge. They brought with them a small skiff powered by an outboard engine, and a fiberglass dory, or canoe. The day had started out calm and sunny, and the crew set off in both boats to catch what fish they could find in this legendary lagoon. A younger nephew in his early twenties, who had grown up in the city and had come along for fun, jumped into the fiberglass canoe and started to settle into the middle seat. Elrick, a cautious person by nature, asked him why he chose the dory. Did he know how to swim? The nephew said no, and Elrick told him to get out of the dory and step into the skiff—a much more stable boat that was also outfitted with life preservers. New River Lagoon is not a place to "play with," in Elrick's words. It turned out that this would be a lifesaving command. Andrew and Ozzie, two well-experienced fishermen and good swimmers ventured off in the dory. Others stayed onshore, or were in the skiff. As the day wore on, the winds began to pick up, at first a little, then a little more, until they were blowing hard and three- to four-foot waves were churning the center of the wide lagoon. Elrick had warned Andrew and Ozzie to be careful—not to go off too far, but they did not follow his advice and instead paddled across the lagoon seeking out the best fishing spots. As the winds picked up, they started to head back across the lagoon, assuming they could make it back to everyone else. Just when some of the others onshore suggested that one truck full of people leave with the skiff, someone shouted that Andrew and Ozzie's dory had broken—split in two—and the two men started to scream for help from the middle of the lagoon. The fiberglass dory had sunk.[1]

The men had no life jackets, no boat to hold on to. Luckily, there had been two plastic gallon jugs (for drinking water) in the dory and each man was able to grab one to help stay afloat. They took off their shoes and clothes so they would not be weighed down and with that went truck keys, wallets and licenses. They were trying to swim ashore—but shore was far. They were lucky that there was a skiff with a motor nearby. Elrick sped the boat to them as fast as he could, and pulled them out of the water—they were nearly hypothermic, and exhausted, but alive. Had the younger nephew been in the dory, he would likely have drowned. After hot-wiring Ozzie's truck (because the key was lost), the group rushed home to get Ozzie and Andrew warm and dry. The story is told and retold frequently, whether on a moonlit night with people gathered around, or across a kitchen table—it serves as yet another cautionary tale about the dangers of the landscape of rural Belize. Even for experienced outdoorsmen, the power of this landscape is intense and unpredictable.

This is but one small example of the myriad ways in which life for rural Creole Belizeans is bound up with the lively, agentive more than human, and how knowledge and understanding of this world is essential for well-being. Significantly, in this case, the more than human can swiftly become overpowering. In recent years, scholars have been paying closer attention to how different people live in their worlds. The so-called ontological turn has been hailed by many as an exciting theoretical development. Primarily the work of many white male academics (and a few others) located mostly in the United States and Europe, who build on European philosophical traditions to analyze South American indigenous populations, this "turn" questions the universality of Western ontological premises. These include assumptions about what kinds of things exist and what kinds of relations obtain between entities. Scholars debate how to analyze and describe the different worlds indigenous peoples inhabit (e.g., Descola 2013a, 2013b; De La Cadena 2015; Kelly 2014). An important feature of these indigenous ontologies as described through European frameworks is a decentering of "the human," a recognition that humans are simply one among a panoply of entities. These scholars and debates have created powerful openings toward fresh ways of thinking in this moment of the Anthropocene, when new approaches are so desperately needed. But one danger of this development is that it so easily reproduces what Michel-Rolph Trouillot has called the "savage slot"—the category of difference that brought anthropology into being, and the investigation of which continues only to reproduce a rigid and unequal divide between the West and the rest (Bessire 2014; Trouillot 2003). Trouillot suggests that the world is much better understood as a place where "there is no Other, but multitudes of others who are all others for different reasons, in spite of totalizing narratives" (Trouillot 2003, 27).

Indigenous and feminist scholars, along with other scholars of color, have been articulating similar points for a much longer time than the theorists most often associated with the ontological turn (Sundberg 2014; Todd 2016). These thinkers from non-European standpoints, writing from a range of cultural traditions, and often from the margins of academia, have long argued that things cannot be understood in isolation, that human being emerges out of a panoply of relations with the more than human, and that the worlds inhabited by non-Europeans, or marginalized populations within Anglo/European locations, are enlivened in ways that escape and transcend scientific and social scientific understandings (see, for example, Allen 1992 [1986]; Deloria 1973, 1997). Indeed, these scholars are as likely to deconstruct the "West" as to try to describe "the rest." Jamaican scholar Sylvia Wynter's analysis of the historical development of the concept of "the human" in European thought, Man1 and Man2, shows precisely how varieties of ways of being human have been excluded from European constructions of this seemingly self-evident category (Wynter 2003; see also McKittrick 2015). I concur with these scholars that there are multiple non-indigenous and indigenous *worlds otherwise*, located in multiple places (see Gibson, Rose, and Fincher 2015). This work to decolonize academic and popular understandings of the world also entails recognizing experiential and spiritual ways of knowing as legitimate and worthy (Sendejo 2014).

What does ethnography that fully embraces these different ways of being and knowing look like? I am not entirely sure, but I work toward that goal in this book and most concertedly in this chapter.[2] I find inspiration in the work of Mario Blaser and Juanita Sundberg. Blaser (2010) effectively shows how the worlds that Yshiro people of Paraguay enact are predicated on different ways of being in materiality, time and space. He uses the framework of political ontology to create room for pluriversality, for a multitude of "others" and their worlds. The Yshiro negotiate their everyday worlds in the tropical forest and with each other, tethered to their own histories at the same time that they and other Paraguayans occupy and engage with overlapping worlds of the Paraguayan state and conservation and development initiatives and organizations. All these groupings of persons, things, and ideas enact their worlds into being through everyday living and in articulation with each other. Stories people tell, policies that are written, laws that are enforced, practices that take place might each give one particular worlding more purchase than another at a particular point in time, but nothing is fixed.

Juanita Sundberg forcefully urges theorists to commit to decolonizing, to— "expos(e) the ontological violence authorized by Eurocentric epistemologies both in scholarship and everyday life" (2014, 34), and instead to build on pluriversality. This entails de-centering the Eurocentric worlding that dominates scholarly production.[3] It means unpacking the academic "we," speaking out about the location of Eurocentric knowledge and building on indigenous, black, chicana, feminist, and other historically marginalized forms of knowledge production and scholarship. Sundberg offers the metaphor of *walking with* for how to move forward in this kind of practice, building on Zapatista conceptualizations of walking the world into being, and walking as an important practice in generating the pluriverse. This idea of walking recognizes the importance of movement and action in bringing the world into being. "Walking with" means engaging "*colleagues* in producing worlds" (Sundberg 2014, 41; emphasis added), or sharing experiences as intellectual and political subjects.

I spend time elaborating my train of thought here because my everyday life entails pluriversality in the most mundane of ways. I am continually enacting Western bourgeois ethnoclass worlding, while my husband continually enacts rural Creole worlding, yet our becomings are also always interfering and intermingling with each other. We negotiate across our multiple worlds daily. In our everyday life in the United States, this manifests most powerfully in how we understand ourselves (and our children) to be healthy or sick, how to remain healthy, and how we relate to our family, friends, and pets. But when we are in Belize, as I share below, I become acutely aware of how the Eurocenteric worlding that brings me into being is out-of-place in rural Creole becomings.

Belizeans take part in the agentive worlds around them with knowledge and skill developed over generations to successfully navigate the dangers posed by the more than human with which they live. This know-how also contributes to particular projects of racialization. In these tropical lowlands, rural Creole socionatural assemblages include skill, knowledge, animals, plants, winds, waters, and more, and a wide variety of skin colors, facial physiognomies, and hair types. Rural Creole brownness and blackness assembles with know-how, competency, and belonging in these challenging environments, providing readings of themselves and their ontological being that lift up blackness and reject white racial framing.

Water

Rural Creole people dwell in watery scapes. Some waterways such as Spanish Creek, Black Creek, and Poor Haul Creek are slow-moving, dark, and deep, rising slowly with the wet season, falling slowly with the dry. Others, like Western, Crooked Tree, and Southern Lagoons are wide and shallow—the water spreading impossibly far across the landscape—flooding acres and acres in the wet season that transform to large grassy savannahs and then desiccated plains of mud and dead grass in the driest part of the dry season (see map 1.2).

Although by the time I started doing research in this part of the world, roads and vehicles were the main modes of transport, not too long before, in the mid-twentieth century, boats were the only way to move between communities and to get to Belize City to sell things, buy things, attend high school, and work. The trip by dory is still well remembered by the "old heads" in any rural Creole community, but especially in those communities like Lemonal, Crooked Tree, Rancho Dolores that both were far "off the main" (a long paddle from the Belize Old River) and lacked roads until the mid- to late twentieth century. To travel by dory from Lemonal entailed paddling down Spanish Creek, across Poor Haul Creek to Crooked Tree Lagoon, finding the passageway to Black Creek, paddling down this narrow but deep twisting waterway to where it meets the Belize Old River, and then traveling down this river another forty miles to Belize City. Memories of these trips reverberate around the contemporary experience of water, and boats still transport people in many ways—across rivers where the bridge crossings are far away (e.g., Flowers Bank, May Pen, parts of Lemonal and Rancho Dolores), in times when the road into Crooked Tree is flooded and the road is not an option to get to school or work, and, of course, boats are still frequently used to access areas for fishing, hunting, making plantashe, grazing cattle, and other activities.

Travel by water is central to rural Creole life, and knowing how to paddle a "dory" is a central competency that rural Creole people value.[4] Knowing how to handle (paddle, steer, get in and out of, sit in—all without capsizing) a "cranky," or wobbly, dory is especially valuable. A rural Belizean never knows when they might need to travel in one across a river, creek, or lagoon, and more dories tend to be cranky than not. In these instances, the agentive capacities of both the cranky dory and the unpredictable waterways keep Creole human being on alert.

Water, whether deep, black, and treacherous in the sinuous course of Black Creek, or whipped up into a frenzy of four-foot waves on New River Lagoon, as I described in the opening vignette, is one of the most powerful and unpredictable entities challenging human being in this place. Belize sits in hurricane territory and torrential downpours and flooding occur every year. Rising and falling waters create mud slicks out of dirt roads, carve off chunks of asphalt, and gouge deep ruts and holes in roads. Water makes vehicle and foot travel treacherous. Waves on deep lagoons can readily swamp any small boats that dare to venture out in these conditions.

"Man drop" rain (big heavy rain drops) and wind can lock people inside their houses for days at a time. In the wet season everything becomes wet and steamy, clothes do not dry, and yards become swamps or fields of mud. The big rivers run swift and become torrents of muddy water—treacherous to travel on or across. Even small creeks become swollen, deep, and dangerous. But then, typically starting in February and March, the dry season begins, and it is not long before lack of water, heat, and parched air and soils characterize this place. Miles and miles of flooded savannah that fed grazing animals from deer to cattle now are barren plains where nothing grows and where cattle and horses struggle to find sufficient grass. The dry hot air concentrates fish in the pockets of

Figure 4.1 Spanish Creek waters run deep, 2017. Photo by Elrick Bonner.

deeper water that remain in receding waterways. Living in this place requires an acute awareness of these shifts, of the oncoming weather, of how likely it is to rain and for how long, if the "dry" has really started, or if there might be another period of heavy rain before it really sets. People become rural Creole through acquiring and putting into practice an ability to read the skies and feel the changes, and knowing the power of wind and rain and sun.

But this competency is being challenged by global climate change. Over the past decade or so, the onsets of both the dry and rainy seasons have been harder to predict and have been happening at unusual times. It has both been hotter and drier for longer than anyone can remember; and the flooding has been more intense than anyone can remember. The worst flooding Crooked Tree has seen in thirty years happened in late 2013 and early 2014—the lagoon rose many feet above its normal height, deeply swamping the "dump," the causeway built across it in 1984. For over three months in 2014, the only way in and out of this large village was by boat, a mode of travel that proved deadly for one villager. When the dry season starts, it seems to be lasting longer, stretching across more months than it should. There is either too much water, or too little and both conditions pose dangers.

Water has power in other ways as well. I was often told that if I drank the water from Spanish Creek in Lemonal, I would never be able to leave this place, I would be permanently "tied," or connected to, the village (see figure 4.1).

Spanish Creek seems to have more power than other waters in this place. I have heard numerous people over the years make cautionary comments to many different people about drinking the water from Spanish Creek. But even the drinking-water wells dug in the limestone formations in the sandy community of Crooked Tree brought occasional comments about how imbibing the water connects one to the place. One day early in my husband's and my courtship, my uncle-in-law-to-be, Uncle Johnny, made me "black tea" from a bucket of water from Spanish Creek.[5] Bodily incorporation of the more than human is here transformative. In my case, it indeed did tie a "wite giahl" (white girl) to "Wreck Bank," Lemonal's affectionate and self-deprecating nickname.

Soil, Mud, Sand, Bushes, and Trees

Winds and muds are also active agents. Hurricane-force winds can rip roofs off houses, haul trees out of the ground, and transform rain into sideway sheets of water. Thick muds slow travel and require one to learn the skill to navigate whether by foot, car, bicycle, or horseback. Trudging through thick, slick mud or across dried rutted landscapes is challenging work. This becomes excruciatingly apparent to me when I am with rural Creole people, and inevitably slow them down because I do not have the embodied knowledge and skill that they do. Walking through any kind of bush also brings the possibility of being cut, sometimes deeply, by a whole range of plants in different ecological settings. The plants of subtropical forests often have sharp thorns and poisonous saps. Pokenoboy palm trees have long, super-sharp, thin spines, not very easy to see but very painful to touch, or the ping-wing of Creole folk song fame (see chapter 5), also has sharp thorns, as do acacia and calabash trees. Other trees and shrubs can cause itchy or painful rashes. The natural environment is replete with plant dangers.

The power of plants can be overwhelming—grasses and shrubby plants can take over human worlds. The subtropical flora in this part of Belize grows very quickly, turning into thick bush in a matter of weeks. As I described earlier, keeping the bush at bay in rural Creole yards is a constant struggle, and occupies time and energy. But while plant life, winds, and water can all be difficult, animals may pose the greatest challenges—their canniness and their sapience causing trouble.

Sharing the World with Other Animals

People become rural Creole through learning about the power and intelligence of the more-than-human creatures with whom they share these watery landscapes. More-than-human animals and beings are willful and canny in ways that do not fully correspond to Anglo-European ontologies. Animals are smart—they know what they are doing in the wild, and make calculating decisions. For example, animals know how to cure themselves with medicinal plants; they know which bitters to eat for what ails them. They know to move to certain places at certain times to avoid being flooded out or burned up. And animals and fish know where and when to move to get out of the way of human predation. They move "to the back," into the less developed areas close to the Guatemala border, when the pressure near Creole villages gets too great.[6] Animals and fish are getting smarter, in this long-term game of tag that the animals and fish play with the hunters and fishers of rural Creole Belize. The animals are always learning more and better ways to outsmart their predators.

Nonhuman animals are powerful as well as smart, and able to transform local ecologies (Haenn 1999; Knight 2000). For instance, people in Crooked Tree, a village famous for its white sandy roads, describe how cattle have changed the roads in the village. Because cattle roam freely throughout the village, cattle manure has intermingled with the pine ridge sand creating soils that are less sandy, where more weeds and bush grow. Wild animals also have the capacity to alter landscapes. Droves of peccary trample and destroy large swaths of shrub and grassland and "mountain cows" (tapir) trample and ruin corn fields and plantashe (cultivated areas) full of ground food. In 2008, one older Creole man described to me how "mountain cow wa bruk up your cornfield, mash ahn all up . . . peccary wa eat out all your groundfood, like cocoyam" (mountain cow will break up your cornfield, mash it all up, peccary will eat out all of the tubers, like cocoyam, you planted). Nonhuman animals have a powerful capacity to transform both natural landscapes and human-made landscapes.[7]

The socionatural entanglement through which rural Creole people come into being is hazardous, and some of the most serious hazards are nonhuman animals. Several particularly important nonhuman animal residents of central Belize remind rural Creole Belizeans, or anyone who lives among them, that humans are not the only powerful creatures in the world. These include snakes like "tommy goffs" (fer-de-lance), rattlesnakes, and "wowla" (boa constrictor), arachnids like "tri-antelopes" (tarantulas), and scorpions. Tommy goffs, rattlesnakes, and tri-antelopes routinely appear in yards, no matter how well kept they are. Where I lived in Crooked Tree in the mid-1990s was no exception and, like many yards, was often full of children, dogs, cats, and chickens. Each occurrence of a snake or scorpion caused concern and fear in children; older teens or adults were called in to remove or destroy the animal. Fear of these animals, and a sense of the power of the natural world, is underpinned by the knowledge that in the past each of these species has inflicted illness or death on a resident of the community. Knowledge of these creatures is ingrained at a very young age, as children move through yards and nearby roadways. The most feared, and the most deadly, is the tommy goff, or fer-de-lance, a relatively common poisonous snake. Although snakebites are not an everyday occurrence, they do happen, and the tommy goff is typically the culprit. Early in 2014, a young woman simply walking along a well-used footpath in Lemonal village was bitten by a tommy goff. She managed to find transport for the hour-long drive to Belize City, received proper medical treatment, and recovered well—only a photo posted on Facebook of her very swollen foot and ankle, and the memory, remain. These stories and memories that come from these experiences are shared and reshared, and generate community-wide understandings of the ever-present danger of snakes and tarantulas and scorpions

A creature that looms large in rural Creole socionatural assemblages is the "tigah." Tigahs include both the "leppaad tigah" (jaguar), which is what rural Creole people usually mean when they use the word "tigah" and the puma or "red tigah." Jaguars and pumas are perhaps the ultimate in willful, powerful, and dangerous animals. They are known for sneakily watching and following people and for their sharp intelligence. Tigahs have a well-established history of preying on dogs, sneaking into villages in the gathering darkness of twilight and taking these beloved and closest of human companions. They occasionally prey on calves and other weak members of herds of cattle, which often roam free in the large expanses of pine savannah skirting rural Creole villages. Tigahs engender the greatest fear when they seem unafraid of people. For instance, in 2002 in the village of Lemonal, two young calves were tied in the yard of an unoccupied house on the outskirts of the village. Barely fifty yards away was a house full of people—adults and children. In the middle of the night one of the calves was silently and stealthily attacked and dragged away by a tigah, the tracks in the morning a testament to the predator's work. Although calves have been killed by tigahs periodically over the years, this encounter was particularly frightening and the story told and retold for months afterward—no one in the neighboring house had heard a peep of noise.

Many people have some kind of story of an encounter with a tigah. Certainly every man who has spent any time in the bush has come across tigahs, and if they were carrying a gun, tried to kill the animal. Women and children who do not travel out to hunt also encounter tigahs, most commonly on roads leading out of villages. Despite all these memories of encounters, I have heard of no one who has actually been attacked by a tigah anywhere or anytime in Belize.[8] But the fear of such an attack is palpable. Many people of all kinds are reluctant to venture into the bush alone, or even to walk the roads between villages or connecting villages to the highway alone, at least partially for fear of attack by tigahs.[9]

The depth of this belief, and my own difficulties living pluriversally, was made very clear to me during my visit to Lemonal in 2004. During the two months I spent in my husband's family's house and yard in the village of Lemonal with my children, husband, and other family members, talk about tigahs was common. Many of the stories told in the evenings were about the tigahs that seemed so particularly plentiful this year—both leppahd and red tigahs had been seen and heard by various people very close to the village in recent weeks. I have longed to see one of these magnificent creatures alive outside of a zoo. I have rarely even had the opportunity to see their tracks or other signs of their presence, and the only one (a jaguar) I have ever seen close-up outside a zoo had been killed and skinned.

One beautiful sunny morning during that 2004 visit, I walked down one of the dirt roads leading out of Lemonal village to the bush. I had planned to video my husband Elrick and his relatives working on building a temporary but large cattle pen on the road a little way out from the village center, which I did indeed do, but the beauty of the day and the open road beckoned. What might I find past the curve? Or over the hill? I slipped away from the cattle pen and the crowd working on it, and headed north on the deeply rutted dirt road into the bush and pine ridge that abuts the New River Lagoon and leads to Crooked Tree. I had been out there a number of times in the past ten years—walking with a group of people, on horseback with friends, in a truck overloaded with people to go out to a swimming hole—I knew what to expect, I thought. The road passes by the newly built primary school where tigah sightings had been plentiful and a missing dog had recently been further testament to their presence. It then cuts through a stretch of high ridge, an elongated copse of tall, densely packed tropical trees, and finally flattens out through a broad grassy palmetto-studded savannah, with the lure of another ridge of high bush far in the distance. I enjoyed walking, enjoyed being alone in the Belizean bush—listening to birds chatter, to the sound of animals (likely lizards) scurrying away at the sound of my (softly, I thought) approaching footsteps. I felt a strange nervousness as I walked past the school, and then again as I walked through a section of high ridge—jaguar territory. I had walked almost a mile and was very much enjoying my solitude. I was alone-in-nature—in a sierra club poster-perfect wilderness landscape, which, although invisible to me, I knew was full of jaguars, ocelots, tapirs, toucans, parrots, and all kinds of tropical wildlife. Just as I was trying to decide whether or not to continue on to the high bush in the distance, I heard the roar of the straining engine of a four-wheel-drive vehicle as it struggled to race along the deeply rutted road toward me. It was Elrick, my life partner of over ten years at this point. He pulled up, screaming, "Get in!" as if we were under siege. I was furious at him for disturbing my solitude and nearly continued on my solo journey out into the bush, but something in me recognized that this might not be very smart. I jumped in, and he quickly turned the truck around and headed back to the safety of the village. We were both too angry to speak to one another. Later he told me he could not understand why I was so willing to risk my life—how could I have gone out into the bush without a companion or a gun—I was acting like a crazy person. I was unable to understand how he could not respect my need for solitude, my love of "communing with nature." I started spouting, "No one has ever been attacked by a jaguar in Belize." He looked at me as if I were the stupidest person he had ever seen. Of course, it finally dawned on me: no one has ever been attacked because no one goes out into the bush alone or without a gun, especially in times and places where tigahs are plentiful. It took me several days to fully understand this very simple point (and several days for Elrick to recover from the fright I had given him).[10] At this moment my particular out-of-placeness was made very clear to me. My white, professional, middle-class, suburban-New-Jersey-born-and-bred self, at a very deep level could not (and still cannot) bodily experience the same sense of power and agency of the natural world as can my rural Belizean Creole husband.

A few days later, to placate my desire to travel in the bush and see as much as possible of tigahs, Elrick loaded up our vehicle with as many people as it could hold. We traveled out through a maze of rutted and swampy dirt roads toward the New River Lagoon, stopping on the way at various points reported by hunters and cattlemen to be likely to have tigah tracks. We found plenty. Only a hundred yards or so from where I had been walking earlier that week there were tigah prints everywhere. The most common tracks were of a small group—probably a mother and her cubs, and some of these were from earlier that morning. My sons, who were four and seven at the time, had fun finding the tracks, and measuring the size of their paws against their hands (see figure 4.2). But none of us ventured far from the vehicle.

An encounter from the mid-1990s with a vibrant young woman, Anna, in Crooked Tree reveals the ways in which tigahs are understood and valued in rural Creole Belize. In the course of an interview about her thoughts and feelings regarding conservation and tourism, Anna got a delighted look on her face and asked me if I wanted to see something really special. We had been sitting in her yard; she went into her house and came back out with a small object enclosed in her hand, which she opened as she came close to me—"a tigah tooth," she smiled hugely, showing me the long curved canine tooth of a jaguar. Anna's husband had killed one on a hunting trip late at night earlier that month, and had returned the next day to salvage what he could from the animal (the skin could be sold for some cash), and he gave his wife one of the teeth.[11] She recounted the story with gusto, admiring her husband's hunting skill, glad the dangerous beast was dead, but also reveling in the animal's sheer physical beauty and power.

These encounters with tigahs are central components of the socionatural assemblages through which people become rural Creole. Apprehending the power and danger of these animals, handling encounters with them, and in some cases killing them, create rural Creole realities and identities. But it is not only tigahs that become entangled with the human in rural Creole socionature.

Another highly dangerous animal is the "halligata," or Morelet's crocodile Notorious among Belizeans for their ferociousness, halligata are increasingly

Figure 4.2 Tigah pawprint, Elrick Jr.'s hand, author's shadow, near New River Lagoon, 2004. Photo by Melissa Johnson.

common in the lagoons, creeks, and rivers of central Belize as a result of conservation policy (see chapter 6). Swimming in lagoon and creek waters is considered dangerous: people rarely swim alone, and they are careful in dories or on riverbanks to make certain no halligata are lurking. I have rarely traveled through waterways in this part of Belize without seeing halligata, sometimes a small one basking on a tree limb in the sun, occasionally groups of them on mudflats, sometimes a huge one swimming. Over the past twenty-five years, I have seen the numbers and sizes of the halligata I encounter increase significantly. Relatedly, during the past fifteen years, a number of fatal and near-fatal encounters between people and halligata have occurred in various places throughout Belize as the once-endangered species has made a comeback. When these attacks occur, they are the center of conversation not only in rural areas, but country wide.

In 2012, a thirty-year-old man eager to earn some cash was diving in the lagoon waters of Crooked Tree and fishing tilapia with a speargun. He likely swam into a halligata's territory and was severely mauled on his head and side. Below is the account of what happened as told by his mother to a reporter for Channel 5 News:

> Well they tell me that when they mi deh together in the waata ih say that ih hear like fish beat; heart beat and ih wah go si how the fish deh stand and

when ih gone deh nuh si ah again so they say deh have to guh look fi ah and when they gone deh nuh si ah and fi wah lee while deh si like wah lee bubbles di come up and when they look deh si ih head and when deh pick up ah ih partly done dead. They had to pump ah and dah suh ih come up back.

(Well, they [the two other young men fishing with the attack victim] tell me that when they were all together in the water, [the victim, Devin] said that he heard a sound like fish beating [a heartbeat-like sound], and that he wanted to go see how the fish were/what they were doing/if there were fish, and when he went, they didn't see him again, so they said they would have to go look for him, and when they went there, they didn't see him and for a little while they saw like some little bubbles were coming up, and when they looked they saw his head, and when they picked him up, he was partly dead; they had to pump him and that was how he came back.)
(http://edition.channel5belize.com/archives/73508)

The crocodile had bitten him on his head, forehead, and shoulder. His friends saw the crocodile, which was huge, about fourteen feet long. The friends who were fishing with him resuscitated him and rushed him to the Belize City Hospital—an hour's drive away. He was immediately put into an induced coma, and his outlook was not clear. Miraculously, he fully recovered within a few weeks, and the next year he was back in the water striking tilapia. His two friends were not so sanguine, saying that they would not dive and strike again, but would instead use nets to fish. This was the first mauling of a person by a crocodile in Crooked Tree in anyone's memory, but it underscored a widespread fear and made clear how powerful these creatures are.

In April 2014, a fisherman in Lord's Bank, the rural Creole village located closest to Belize City, but still connected by kin and friendship networks to the other villages in the Lower Belize River Valley, was mauled and killed by a ten-foot halligata. The fisherman was wading in a man-made pond when children saw a halligata drag him in. The creature was either a Morelet's crocodile or the American crocodile, which prefers more brackish water, as might be found closer to the mouth of the Belize River; in either case a deadly beast. It took a search party hours to find the body, and when they did, the man's right leg had been nearly severed off. These incidents serve as visceral reminders of the dire dangers of "life da bush" and the power of the animals that live there.

This tension between humans and crocodilians, reptiles that have the potential to turn humans into prey, has been carefully considered by Val Plumwood (Plumwood 2012; see also Quammen 2004). Plumwood, a pre-eminent environmental philosopher, herself survived an attack by a crocodile in her home country of Australia. Thinking through her experience of having become prey, Plumwood deconstructs Eurocentric understandings of humans as masters of the natural world. In a different consideration of crocodilians, Laura Ogden's beautiful description of the difficult work that Florida's alligator hunters engaged in to eke out a living in the Everglades in the early twentieth century also highlights the vulnerability of the human and the power of the alligator (Ogden 2011). Both accounts reveal a reality well grasped by rural Belizeans: humans are merely one among many of the forms of life that inhabit the planet, and many of those other forms pose mortal threats to human being.

Other Beings

Being rural Creole also means potentially encountering other kinds of powerful and canny creatures that may or may not be material (Craig 1991). The most widely talked about and commonly encountered of these is the dwarf, "Tata Duende."[12] Tata Duende is a spirit creature of which children are especially afraid, though very few people (myself included) enter the bush without some thought about the presence of Tata. Tata, a short dwarf, has feet that face backward and is missing his thumbs. He sports a large hat and often walks with a knotted stick. He steals children who wander into the forest, especially if they show their thumbs, which he interprets as their making fun of him. I have heard many firsthand accounts of sightings of Tata, and relate a few of these below. One woman shared her story of being tricked by Tata when she was in her early twenties (this would have been in the early 1990s). Tata led her out into the pine ridge outside of Crooked Tree where she got lost and spent two days wandering until she recognized where she was and was able to find her way home. Tata can shape shift as well. An older man, Alpheus, told me of his encounter with Tata when he was a young man hunting in the New River Pine Ridge with his father. He saw an extraordinarily large currasow, and began to raise his gun to shoot it, when his father shouted at him to stop, having seen a man from a neighboring village, but when his father called out to this man whom he knew, the man was not there. As Alpheus put his gun down, Tata resumed his dwarf shape and disappeared. Alpheus and his father had both been tricked by the dwarf. In another instance, Isobel, a woman from Lemonal who was very skilled in navigating the bush, was out with two of her children, and was led away from the village for days by Tata and his tricks. When she was finally found she was nearly ten miles from Lemonal and close to a neighboring village. In another instance, a cattleman from another village, Maypen, was out with his family several miles from the village at his plantashe, planting rice. The family had brought their baby who could not yet walk with them, and had put the baby at the edge of the area where they were working. When they went back to check on the child, he was gone. Teams of searchers, police, and dogs hunted for days and could not find the child; the only plausible explanation was that Tata Duende must have taken him.

Because Tata is "devilments," most people believe he cannot be killed. Hector's story illustrates this. When he was a young man, Hector went out hunting with his father. They were not having much success. Hector decided to try his luck a distance away from his father. Hector's father, knowing his son was a good bush-man, left for home. As evening came on, Hector saw something sizable move and shot at it; from the sound and movement that followed he knew he had aimed accurately. But then he heard all kinds of bawling and unearthly noise coming toward him and climbed a tree for safety—below him was the dwarf, fully alive, with no evidence of blood or injury of any kind. All night the angry Tata circled the tree below, bawling. The creature finally left just before morning, and Hector made his way home, fast. These are but a few of the plentiful stories of encounters with Tata Duende that circulate in rural Creole communities and that serve to remind people of the range of difficulties, including death, that Tata can cause.

But Tata is not the only devilment who lives in the bush. The "Jack-o-Lantern" is commonly encountered in rural Belize (for sightings in South America, see Escolar 2012). This entity is a golden orb that hovers above the earth out in the pine savannah, and while not necessarily dangerous, is considered to be the work of the devil and unpleasant to encounter. Almost every person who hunts at night has come across one. My husband and his older cousin witnessed Jack-o-Lantern firsthand when Elrick was in his twenties. They watched the orb start as a small red dot just above the ground in the distance, darting along. It became larger and came closer, close enough that they could see how it lit up the ground below. It then sped by them and condensed back to a small red orb, as if beckoning them to follow it. They chose not to.

There are other less frequently talked about immaterial, or materially different, creatures. Hunters might encounter "wari massa," an enormous and super-powerful wari (a type of peccary) that kills to protect its drove of fellow waris. Wari massa protects its drove from jaguars and humans alike. If a hunter shoots wari when the wari massa is nearby, the wari massa will scream and the hunter had best find a tree to escape to as quickly as possible. "Ashishi pompi" are harmless tiny dwarflike creatures that like to play in the ashes of a fire hearth or an old fire. If they are present, one can watch them play as the fire dies down, but they are hard to see because they are the same color as the ashes. "Old heg" (old hag) is a ghostly female vampiric figure, who slowly sucks the life out of children. And everywhere are ghosts. Everyone knows what places are likely to have ghosts and most people prefer not to pass these areas after dark, and if they have to, they go fast. One place, Wade Bank, between Crooked Tree and Lemonal on Spanish Creek, the site for several years of a fishing lodge for U.S. tourists, has long been home to an Indian woman ghost, reflecting the long and widespread presence of Maya in the lower Belize River Valley.

Conclusion

This place is fraught with danger and unpredictability, from only partially material entities to large predators and swift waters. Knowing the possible dangers, and knowing how to "handle one's candle," as the Belize Kriol phrase goes, is essential for living well here, and is also a defining characteristic of being rural Creole. One becomes rural Creole through being able to navigate this agentive world.

The ontological separation between human and more than human and between corporeal and spirit does not hold as tightly or fall quite along the same lines in rural Creole communities as it does in Anglo- and Eurocentric scientific and popular thought. Humans do not hold the full deck of cards, while nature awaits the next human impingement—the story is as likely to be the other way around. The bush is unpredictable; it is a dangerous place filled with canny creatures. The worlds otherwise in which Belizean Creoles live already de-center the human in ways that scholars today are suggesting is critical. Human supremacy is far from taken for granted, and many elements of the more than human occupy a willful and cognizant ontological status, just like the human.

Dwelling in this type of world brings rural Creole Belizeans into being. Living well in this powerful landscape requires a great deal of skill and knowledge. If the colonial racial order relegated nonwhite and poor white bodies to these marginal lowlands, the depth of skill and knowledge rural Creole people have developed and passed along to their descendants over the years has generated an association of strength, intelligence, know-how, and competence with being Creole

CHAPTER 5

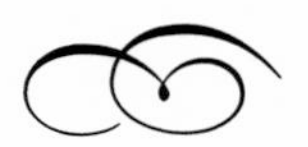

Entangling the More than Human

BECOMING CREOLE

The sun had not yet risen, yet I was up and out of bed as early as Mr. R. and Miss L. for the first time since I had moved into their Crooked Tree home in the early 1990s. I hastily pulled on long protective pants, threw some clothes into a satchel, and ran off to meet Paul and Rodney, two brothers admired in the community for both their love of and skill at hunting, who had promised to take me on a tigah hunt. I longed to see a tigah, or jaguar, in the wild, and to participate in an overnight hunting expedition in the New River Pine Ridge. I entered their yard to find everyone still asleep . . . had I gotten the day wrong? They said to be there by 5:00. . . . Feeling silly, I stood around for a few minutes. The dogs began to bark, and at this commotion, Paul emerged from his sleeping house and Rodney emerged from his a few moments later. They gathered together their equipment—guns, machetes, sacks, dogs (coonhounds—two of which would produce a puppy who was my companion for many years on walks through the wilds of U.S. suburban neighborhoods in Michigan and Texas, where the scent of ever-present deer never stopped tantalizing him)—and we headed off to Georgie Pass on Western Lagoon. When we reached the lagoon edge, we loaded our gear, dogs, and selves into a dory, and Paul and Rodney poled and paddled the long way out through the logwood swamp, pointing out, to our mutual amusement, bushes like Old Lady Stinking Toe and Juke-Mi-Back along the way. We finally reached the open lagoon and swiftly crossed to the other side. The path up out of the lagoon lowlands was steep and slick with the blackest of cohune mud; I barely made it up. Paul waited for me, Rodney was long out of sight. As soon as we reached their father's camp in the pine ridge, the two men took off on horseback to catch some "dinna." I sat alone, enjoying the beauty of my surroundings, the long vista across waving plains of grass, punctuated by differently shaped and wooded glades, with high ridge in the distance, wondering if they would expect me to drink water out of the black-water pond near the camp, wondering what they would bring back. I did not wait long. They caught two "hamadilly," armadillos, and Paul cleaned and cooked them—one for us, one for the dogs. Rodney took me out on horseback to where he had taken some ecotourists a few weeks earlier to see some jabiru storks in their nest high in a pine tree. As we were riding, Rodney told me how much he loved it out here, how he would rather come spend the night at the camp than go to a dance at the club any day. I thought, how interesting, this sounds like one of my Greenpeace friends talking during a backpacking trip to Yosemite. The jabiru were not there, the skies turned leaden, and there were streaks of lightning in the sky. I became frightened, remembering the stories of rural Creole people who had been struck by lightning and killed or permanently scarred. Rodney laughed, but then took us back to the camp. Back at camp, Rodney shook out the bedclothes and checked under the mattresses, where a long fat tommy-goff (fer-de-lance, one of the deadliest snakes in the world), was sleeping, coiled under Rodney's mattress. With one swift slice of the machete, Rodney cut off his head and we watched the dying snake slither and uncoil. As darkness came, Paul turned on the radio, raising the volume high and I tried to fall asleep to the pounding rhythms of soca and reggae. The sun woke us all early the next morning and we headed out for a full day's hunt. During the course of six hours, with the expert help of their dogs, and despite my clumsiness on horseback, Paul and Rodney caught a gibnut, a deer, and three more hamadilly—six hours yielded plenty to share with their mother, who was well known for her excellent cooking, and other relatives and friends. I watched them skin and butcher each animal, draining the blood before we traveled and making sure their dogs only ate certain parts of each one. This was potentially a very messy process, especially for the large deer, but the brothers were neat and made the task seem easy. Along the way, they stopped and showed me tracks of the famed tigah, smaller tigah cats (margays and ocelots), hamadilly, and gibnut. They explained the behavior patterns of the animals they were hunting and pointed out places that I could not tell apart, noting how one place had "plenty gibnut," another "plenty hamadilly," and yet another would be a likely place to find a sleeping tigah. At the day's end, we reversed our long trek through the pine ridge and swamps, reaching the village well after the sun had set.

The sense of cultural identity and community belonging that mark what it is to be rural Creole emerges from relationship between humans and the more than human in this place. The intimate, intentional, and skilled engagement rural Creole people have with the more than human around them brings these Belizeans into being. These relationships themselves assemble with global political, economic, and cultural processes that shape the institutions, practices, and meanings that contribute to rural Creole becoming. Yet one of the central arguments of this book is that rural Creole people, in this place (and the other places they live), cannot be understood first and foremost as creations of global process. Rather, they are active creators of their own ecologies and economies. Through their everyday activity, enskilment, and practical engagement with this place, they craft rural Creole selves in ways that are not fully or even primarily a product of capitalism's political and economic structures. This argument follows the thinking of feminist political economists J. K. Gibson-Graham, who exhort scholars to see the full variety economic practices that people engage in (Gibson-Graham 2006a, 2006b, 2008, 2011).[1] They claim that scholars inaccurately imagined capitalism as overdetermining economic life. They suggest that there are many examples of noncapitalist economic forms and practices around that world that should be identified and described. I am sympathetic to this viewpoint. Indeed, when I first encountered Gibson-Graham, it was as if a lightbulb went off; this was a productive way to frame so much of the economic activity I had witnessed over the years. But I also want to extend Gibson-Graham's logic to thinking about human ecological relationships. A great deal of the scholarship in environmental anthropology, geography and related disciplines overemphasizes the power of neoliberal thought and governance to determine local environmental activity (but see, e.g., Cepek 2011; Hathaway 2013).

That said, in important ways, the racialized political economy has shaped and continues to shape life here in rural Belize. Racialized political and economic processes and formations encouraged less elite free brown and black people to settle "off the main" in the eighteenth century, and these links between race and place limited opportunities for capital accumulation and development for poorer and darker-skinned people. But I also do not want to *assume* that the people who settled in the lower Belize River Valley would necessarily have *chosen* to partake in the project of capitalist accumulation had they been able to. The original settlers of places like Crooked Tree, Lemonal, May Pen, and Flowers Bank may have been excluded from living elsewhere, but, we also cannot know for certain what they wanted to do, where and how they wanted to live. Assuming that these individuals would have chosen to accumulate capital is a manifestation of what Sylvia Wynter calls the over-representation of Man2, or having *homo oeconomicus* as the stand-in for all people.

My analysis builds on scholarship of and from the Caribbean (and the Americas more broadly) on Afro-Caribbean and American ecocultures *otherwise* that developed in the margins of the plantation complex that dominated the region. Michaeline Crichlow and Patricia Northover (2009a) put forward a productive new theory of creolization, arguing that Afro-Caribbean peoples were "entangled with emancipatory projects" (p. 19) through place-making on the margins of plantation ecologies and economies. Here, Afro-Caribbean peoples were creative agents in dynamic processes of crafting something new from their African, European, indigenous, and other ancestries. Similarly, Monique Allewaert (2013), in her exploration of personhood in colonial tropical Afro-America, argues that alternative ecologies were brought into being through the intimate relationship between Africans and African-descended peoples and the more-than-human world around them in the interstices of lowland plantation society, often in the swampier parts that could not be readily yoked into industrial agricultural production. The personhood generated through these processes is one that emerges out of these ecological relations. Relational, or ecological, ways of being not only occurred during the historical moments of colonial and plantation tropical America, but are present today as well throughout the Caribbean. In this chapter, I describe the relational mode of being, the being *otherwise*, that persists in rural Belize in the twenty-first century.

Tim Ingold's concepts of dwelling, enskilment, and practical engagement are useful in making sense of rural Belizean socionatural entanglements (Ingold 2000). Humans are always participating in embodied engagement with the world around them. As humans practically engage with the more than human and share these experiences and the knowledge they generate, they become enskilled, in this case as hunters, fisherfolk, cattlemen, cooks, navigators of difficult landscapes, and more. Out of these encounters emerge culture and identity: we become who we are by dwelling in the world. Rural Creole people, through repeated practical engagement and enskilment in living in these landscapes, create relational ecologies and economics and become rural Creole. While larger political economies are always also present, the forms of ecological and economic activity that most matter to being rural Creole are those that are *otherwise*. Furthermore, these ecologies and economies not only are represented in the material connection between the human and the more than human but also are central to rural Creole cultural production, from language play and use of metaphor to making music.

Becoming Rural Creole: Practical Engagement with the More than Human

As described earlier, knowing how to navigate through a powerful landscape that is replete with mortal danger to humans in the form of waters, winds, snakes, big cats, reptiles, and indifferent and malevolent spirits is an essential part of being rural Creole. Equally important to becoming rural Creole is their everyday practical engagement with the materiality of the landscape around them. Being rural Creole emerges from an enskilment that entails knowing how to eke out a living from these landscapes in a way that typically melds wage labor with what Belizeans call "kech and kill." Kech and kill refers to a wide range of subsistence activities, none of which pay well, but cobbled together, can be enough to sustain a person and also allow some freedom to do things in their own time. The Caribbean has long been characterized by this approach to economic activity. Lambros Comitas coined the term "occupational multiplicity" to describe it (Comitas 1973). In Puerto Rico this strategy is called "chiripas" (Griffith and Valdés Pizzini 2002), in Jamaica, "making do" (Bolles 1996; Ulysse 2007), in Guyana, "making a life" (Williams 1991).[2] Caribbean peoples are experts at putting together a variety of activities to sustain themselves and do so in ways that allow them some control over their lives.

In rural Belize, many of these kech-and-kill activities are just that: catching and killing, directly tied to an engagement with the more than human around them. Knowing how to use plants and animals and doing it well are badges of rural Creole pride. The specific activities that are most important at particular historical moments partially reflect national and transnational political and economic trends such as population growth and decline through immigration and emigration, shifts in the types of wage labor opportunities available, and the presence or absence of markets for products such as crocodile and jaguar skins. Yet some activities—fishing, hunting, making plantashe—persist through economic and political vicissitudes. Fishing might be the most important activity that contributes to becoming rural Creole.

Fish

I rarely meet a rural Belizean who does not love fish. Rural Creole Belizeans love their shallow-water river fish—crana, bay snook, tuba, and now even tilapia (see chapter 7), and the deepwater tarpon. Catching fish, cooking fish, and eating fish are important elements of the socionatural assemblage that is rural Creole culture. Villages are located adjacent to wide lagoons, slow, deep creeks, and fast rivers, places that attract large schools of many different species of fish—each species doing their world-making in particular spots and habitats. Crana, bay snook, tuba, and others are common in the lagoons near Lemonal and Crooked Tree. Indeed, crana serves as a brand or symbol of Crooked Tree. If you tell someone you are going to Crooked Tree—they say "you gwine eat up wah latta crana dehn?" (so, you will eat up a lot of crana?).

During a visit to Lemonal in June 2013, Mary, a quiet middle-aged mother with bright eyes and a ready chuckle, encouraged me to join her husband, Evan—one of a few rural Creole individuals who, at that time, fished as a main way of making a living—on an afternoon fishing trip. Also planning to go were her brother-in-law Steve, and my husband, Elrick. Mary's son EJ, and my two teenage sons joined the expedition. We drove off in Elrick's pickup truck, taking the very rough dirt road that leads out northwestwardly from the village, toward the southern edges of New River Lagoon, where ponds were still likely dry enough to harbor dense congregations of fish. We traveled through pine savannah, and then the road cut through a small piece of high ridge, until the ground started to soften and mud threatened to trap the truck. Evan jumped out of the truck back with his large nets, his brother-in-law followed right after, carrying a bicycle to ease the long trip back if they would be lucky enough to have a large catch. From here we walked a mile on a muddy track, through areas that had been underwater several months before, until we reached a small, shallow pond near the lagoon. After carefully scanning the water, Evan chose a spot to enter and then moved slowly through the shallow water, dragging the huge net behind him, and finding the sandier spots to hold his footing, not the thick mud that would drag his foot down and be hard to move out of. The fish would move away from him when they could, if they sensed him in time. His goal was to move at just the right speed and direction to capture as many as possible. Evan knew this water and land like the back of his hand. The only thing he could not predict was whether or not the fish would be plentiful that day, but he thought he might get lucky. The nets were huge, the water shallow, wide, and grassy, the bottom muddy. He was right, the catch was plentiful, mostly tilapia, but lots of bay snook and tuba, too. He strained to pull the net out of the water, but the company he had brought with him today made it easier to drag the net along the soft ground with its hundred pounds of fish to where the bicycle had been left. He heaped the net and fish onto the bicycle and he and his brother-in-law pushed the bicycle with its heavy load along the rutted muddy pathway until they reached the truck.

Fishing is hard work that mingles human and fish bodies and water and mud and sand. It takes immense patience, care and knowledge—this is practical engagement accompanied by the required enskilment. A fishing expedition like this re-inforces Evan's beingness as a rural Belizean Creole fisherman. He becomes rural Creole through putting into practice his knowledge of the many creeks, ponds, and lagoons in the area, of the types of nets most useful in the dry season, of whether line fishing or striking (using a spear) would be the best technique for the place, of which technology to use for which type of fish, or how best to fish whether the water is high or low. Yet it is not only Evan, a renowned fisherman, who replenishes and renews being rural Creole through this experience. Everyone who joined the trip also had dimensions of who they were re-created and re-inforced. I was reminded of my deep lack of competence in this place, my limited abilities and knowledge that were made most clear in my only being able to slowly pick my way along the slippery, muddy road. Mary, on the other hand, while clearly capable of keeping up with her fleet-footed ten-year-old son EJ who was eager to catch up with his father on the bike, waited for me, ensuring my welfare. Her competence was underscored by her kind act of caretaking of the white lady. Elrick and Steve, although not as deeply connected, as intimately familiar with this particular place, or with the craft of fishing as a main source of income as Evan was, were ready to offer their still rich fishing and hunting knowledge and enskilment to Evan. They put into practice, and thereby renewed and replenished, their ability to navigate life da bush on the edges of lagoons and creeks.

We returned to the village shortly before dark. Elrick and I were exhausted. Evan and his family hurried home with the catch, and Evan spent much of the night cleaning and filleting the fish. He carefully packed tilapia fillets and whole bay snook and tuba into his large deep freezer, preparing them for delivery to shopkeepers in neighboring villages in the morning. Evan's knowledge and skill allowed him to dispatch the roughly one hundred pounds of fish swiftly. This is good: only in the short-lived dry season can he make such large catches, by contrast the haul is light when the water is high. This year the money he makes from this haul will go to a big celebration—his eldest daughter's graduation from high school, but it will also help the family when catches are lean, very lean, during the months and months of high water in the wet season.

This moment I describe is just one way in which fish, mud, water, trucks, nets, bicycles, sun, and people come together to renew and re-create what it is to be rural Creole. Evan fishes in a wide range of locations, depending on the season, the moon, his mood, what friends ask him for, and what shopkeepers want to buy. He targets a range of fish species, and uses line and hook, spears, and nets of different sizes and types to catch the fish. There are only a few people—both women and men—in each of these villages whose main livelihood is fishing, as it was for Evan. But nearly every Creole person knows how to line fish or spear fish, and many enjoy catching fresh fish for tea (breakfast or supper), or dinner (the main meal of the day). The process of line fishing or striking from a dory or from the bank of a creek or lagoon is enjoyable in itself, at the same time that fish is a favorite food.

Knowing how to prepare the fish to be eaten is also a part of being rural Creole. Fish have to be cleaned—scales scraped off, guts scraped out (see figure 5.1). If one is cleaning a tilapia, the head is cut off—very few rural Creole people want to eat the head of a tilapia or suck out its eye, but the head and eye are the most desired (and often fought over) for native crana, bay snook, and tuba.[3]

The fish can be fried dry and crispy, to go along with flour tortilla or johnnycake, for a "morning tea," or breakfast. Or it may be made into a "sere," steamed with coconut milk and onion, and served with rice for dinner. If the fish is the huge tarpon or a large baca (catfish), a pound or two of flesh can be ground and made into fish balls, a popular treat typically served with white rice. Women are the primary cooks in rural Creole villages, but many men will at different points try their hand at these dishes, and also take pride in

Figure 5.1 Cleaning a bay snook, Lemonal, spring 2017. Photo by Elrick Bonner.

knowing how to prepare them well. And for everyone, *eating fish*, especially these river and lagoon fish, is critical to being Creole: fish grace rural Creole tables frequently.

Turtle

When turtles encounter humans in and alongside the waterways of this part of Belize, rural Creole culture is (re)created. Creole hunters seek out turtles, using nets and spears, diving into dark waters, or reaching into deep mudholes to capture these animals that are a favored food. Turtles, especially sea turtles, historically played a role in the life of the people who made this place their home. A common occupation listed in the earliest censuses of Belize in the late eighteenth century is "turtler." Sea turtles were common and easy to catch in the earliest days of European and African settlement of Belize, but the appeal of turtles spread to inland freshwater settlements as well. One of the most popular wild foods in Belize is the hicatee, a flat-shelled river-dwelling mud turtle. This turtle is highly endangered throughout its range in Central America, but still has relatively healthy populations in Belize.[4] Although the hicatee is by far the most important turtle both culturally and economically, two other species are also caught and either sold or eaten: laagrahed and bocatora.

Hicatee is a prized part of the rural Creole diet, especially during Easter season, when it is traditionally hunted the most. Easter dinner is "hicatee and white rice," and everyone wants to serve hicatee on this holiday. Hicatee are hunted either with a "peg" (spear) during regular fishing trips, with a net over holes where they may have gotten trapped as the high waters recede in the dry season, or by diving in areas where the water is clear, such as the Belize River (Polisar and Horwich 1994; Rainwater et al. 2012).[5] Laagrahed, hunted in similar fashion, are usually caught accidentally when hicatee are being sought. Bocatora, turtles that spend substantial time out of water, are hunted opportunistically: if a bocatora happens to be crossing the path ahead of you, you pick it up and take it home to cook. Sometimes people will organize small bocatora gathering parties and will head to a pond in the dry season that is known to be a bocatora hideout, dig a three-foot-deep hole, and reach into it in just the right way to pull out this occasional treat. As with fish, turtles are a potentially good source of money, especially in the dry season when they congregate in small ponds. Otherwise, people occasionally hunt a hicatee, bocatora, or laagrahed when they see one. Being rural Creole is not only knowing the best method of capturing a turtle but also knowing where the best turtle holes are. Hicatee, especially, settle in particular spots in deep creeks like Black Creek and Irish Creek, and also in the Belize River. Each time a Creole person finds and captures a turtle, knowledge and enskilment are replenished, as is being rural Creole.

People also become rural Creole through knowing that a live caught turtle, stored on its back in a cool and shady place, keeps for weeks. People become rural Creole through knowing how to clean a turtle—how to remove the shell, how to cut the meat and find the eggs. Both men and women pride themselves in washing, seasoning, and cooking the meat to perfection, "stewing" (browning and braising) the meat so that it is soft and "nice," or very tasty.

Deer, Hamadilly, Gibnut, Peccary, and Other More-than-Human Animals

Becoming rural Creole happens not only through interactions with water-bound creatures like fish and river turtle, but also through practical engagement with a wide variety of terrestrial animals. Rural Creole people have always hunted for sale (to brokers, to Belize City markets, to others in the community), but this has also always only been one part of the role that hunting plays in people's lives. Men especially enjoy hunting, enjoy their competence in the bush, finding animals, and then bringing back to the community a treat to be shared and savored.[6] Hunting experiences entangle rural Creole people with the more than human: the fog, heat, mosquitoes, mud, the dogs that run along and chase, the horse, bicycle, truck, ATV, or dory that carries the hunter, and of course the animals the hunter seeks (see figure 5.2).

Figure 5.2 Loading dories with gear for an overnight hunting and fishing trip, Lemonal, 2017. Photo by Elrick Bonner.

Although most hunting is for game meat to either consume or exchange, people also kill animals to protect agricultural fields. For example, a farmer might shoot a "mountain cow" (tapir), because they root up planted areas, protect cattle by removing a tigah, or rid the place of vermin like snakes and halligata (crocodiles).[7] Or some men might kill an animal for the simple satisfaction of aiming and accurately shooting.[8]

The animals most commonly hunted for food are deer, gibnut (paca), peccary, hamadilly (armadillo), Indian rabbit (agouti), wari (another species of peccary), and iguana, sometimes affectionately called bamboo chicken. Certain birds, such as quam (crested guan) and currasow are popular for their meat but difficult to find, and whistling ducks, cluckin hen (limpkin), toby-full-pot (great blue heron), and perhaps even an occasional jabiru have been hunted in the past, but are less so today. Other species are occasionally hunted, but are considered barely acceptable as foods by most people: "quash" (coatimundi), possum, squirrel, and parrot, for example.

In 2013, I sat down to formally interview one of the most skilled hunters in the village of Lemonal, Osmund, or Ozzie. I have known Ozzie for twenty years, and had always assumed that he sold most of what he hunted, either to fellow villagers or to the meat markets in Belize City. I thought that while for him hunting was undoubtedly fun, it was most importantly a source of income. During this interview, I discovered how utterly wrong I had been in my assumptions: he almost never sells what he catches.[9] I knew there were other men in the village who hunted far less frequently than he did, who typically did sell what they caught. But for Ozzie, hunting is most of all simple enjoyment, and if anything more, a way to connect with his friends and relatives in the village by sharing the catch with them. Indeed, other villagers have remarked that he *should* sell what he catches, and not always be so willing to give the meat away. The word that most peppered his speech in our conversation was "freedom," a sentiment that rural Creole people so often evoke about living life da bush. He beamed with pride as he described the best times and places to hunt, and his fondest memories, his earliest experiences. Despite being a naturally quiet person, he was animated and lively during this interview, his confidence in knowing the bush, knowing exactly when, where, and how to hunt shining brightly as he talked.

The hunting experience—the nuts and bolts of it—deeply mingle human and more than human as well as tax the human body. Hunters travel far to find good hunting grounds, often leaving villages well before sunrise. Finding a deer, chasing it, and finally shooting it are difficult enough. But then the deer is cleaned, the guts removed, and most of the blood drained before returning from the hunt. This is a messy, smelly job, and the still dripping carcass is then loaded onto a horse or slung over a shoulder—an intimate relationship between the hunter and the prey. A deer, in particular, is also heavy to "back," or carry—the weight can be exhausting, a horse can ease the struggle, but it is still difficult.

In the mid-2000s, a small group of men went out hunting together from Lemonal and had a particularly successful hunt—four peccary and a gibnut. Once the animals were brought back to the village, each was carefully skinned and cleaned and then cut into smaller portions to be shared with the closest kin and friends of the men participating in the trip. The job of cleaning and butchering the catch was shared with others. One man known for his expertise in cleaning gibnut by keeping just the right amount of fat to make the meat as tasty as possible set to work on the gibnut. Another, particularly skilled in cleaning peccary, which have foul-smelling scent glands that if accidentally cut render inedible the meat of the entire animal, took on the task of processing the four peccaries. They had set up makeshift tables in the center of the village where they processed this food. Other community members and children, as well as dogs on the lookout for discarded bits of meat, all gathered around to keep the men company. The sociality of this day was broad and deep. The intimate connection that the hunters and community members made with the animals they killed and cleaned, and the gathering of many villagers to share in this afternoon's activities on this day entangled the more than human into rural Creole lifeways.

Turning animals into food most typically happens through hunting trips out into the bush beyond village perimeters, but that is not the only way a wild animal can become food. If he kills a mother animal and the baby is close by, a hunter will sometimes bring the baby back to the village, and either raise it himself or give it to a child to raise. The animals most commonly raised in this way and kept as pets are gibnut, deer, and peccary. Once they have reached adulthood, they sometimes continue to be kept as pets for a while, but are eventually slaughtered and eaten.[10] The intimacy here between wild animal and human animal incorporates nurturing, killing, and consuming—perhaps the lines between these ways of being are not as hard and fast as commonly understood in the Global North's dominant culture.[11] Hunting and fishing are powerful ways in which human and (relatively large) more-than-human animals engage with one another, because of the bodily intimacy—the sharing of fluids and flesh. But, people become rural Belizean Creole through a range of other relationships between the human and more than human, as well.

Domesticated More than Human: Cattle and Others

Cattle, and to a lesser extent other livestock species (horses, pigs, goats, sheep) are also important to being rural Creole. Herd sizes range from several cattle to over a hundred head, they are usually Brahman, and are kept mostly in outlying areas, where they range freely.[12] Often, land that had been used for a plantashe for a number of years is burned, cleared, and planted with grass to be used as pastureland for cattle, creating the cohune palm-dotted grasslands that distinctively mark the Belizean landscape. Historically, cattle have been allowed to roam free in the savannah grasslands that border the many small waterways in the areas surrounding villages. In recent years, it is becoming more common to fence in a particular area of pasture for cattle, minimizing theft, predation by jaguars, and other losses. In Crooked Tree, an initiative put forward in 2004 by a former village council chairman (who was also a sanctuary warden at the time) to promote fencing also exhorted the ecological benefits of restraining cattle. Although cattle are occasionally sold to Belize City vendors for cash, most cattle owners use their cows as a bank account—a location of stored value that can come in handy if one needs a bit of cash quick. But cattle are far more than a simple economic investment. They indicate wealth and status in that not everyone can afford to own them. But they also simply occupy a special place in the hearts of the Creole families, and particularly the cattlemen responsible for their welfare (see also Hoelle 2015).

For example, cattleman Mr. Raymond in Crooked Tree, who was in his seventies in 2013, and who "travels with" (has chronic) back pain as a result of many years of hard work, nonetheless still made daily visits to his pasture, about seven miles roundtrip, to check on his cattle. For many years he made this trip on

Figure 5.3 Cleaning a cow, Lemonal, 2012. Photo by Elrick Bonner.

horseback when there was no road. But by 2013 he often used a vehicle to get there, but still went by horse, foot, or dory if necessary. My husband also deeply loves and cares for cattle. He knows each of his cows individually.[13] When one dies, it is mourned. A powerful pull to Belize for my husband is the pleasure of gazing on his cattle, rounding them up to brand the calves, giving them "drench" (or deworming medicine) and otherwise taking care of them.

Rural Creole Belizeans' embodied relationships with these cattle include slaughtering and butchering the animal if the meat will be sold, shared, and used primarily in the village (see figure 5.3). In Lemonal, this task is carried out before a big wake, or for some other large social occasion like a birthday party or wedding. As described above for butchering game, men are known for different skills in the various tasks required, and a crowd inevitably gathers to watch. Children watch closely, learning how to kill, skin, and butcher. These are very good times for the dogs of the community, who wait to be invited to eat the parts left for them at the end of the slaughter. If the cow is to be cooked immediately for a party, the meat is seasoned in large five-gallon buckets (often by a man known for his seasoning abilities), and ideally left to sit overnight; and then stewed—browned and then braised for a long time until the meat is tender. Or it may be readied for barbecuing: braised for a shorter time (usually by a woman) and then barbecued (usually by a man). In the end, a wide range of people have handled the cow, the carcass, the meat, the cooking—all engaging in intimate relationships with this animal, before everyone invited to a party, attending a wake, or however gathered together, enjoys eating it.

On any evening in a rural Creole village, one is likely to see kids chasing each other around a yard—one child with a lasso well looped up, another running in front. They are playing cow, even children (more often boys, but not exclusively) as young as three years old try to throw a rope and catch a cow. As children get a little older they might try to rope a tied-up calf—but this can be dangerous because mother cows are very protective of their young. The older and more mischievous boys are the most likely to try this, despite the danger, and are quickly admonished by any nearby adults—but nevertheless are simultaneously admired for their fearlessness and growing skill.

In addition to cattle, people also keep pigs, sheep, and horses. Because pork is popular at Christmastime, pigs can be a good source of cash at this time of year when people typically need sizable sums to paint their houses, put down new flooring, buy new furniture, and buy food and drink sufficient to serve the crowds of visitors that Christmas inevitably brings. People raise horses for several reasons: for the sheer enjoyment of owning horses, as a practical means of transport in this very variable topography (from swamp to arid plain), to race in the local horse races, and, where there is a demand for it, as an economic investment in ecotourism. Many households raise an assortment of fowl, such as chicken, turkey, duck, and guinea hens, most often for home use but also to sell within the village. In the 2010s, rural yard-raised chickens also became valuable commodities. They are bought by Chinese Belizeans, whose population is growing, and who prefer "local chicken" (chickens raised in people's yards, not on factory farms, which in Belize are run mostly by Mennonites) for their own consumption.

The close connections rural Creole people have with their domesticated animals and birds, caring for them, slaughtering them, and preparing meat to share and use, and cooking them are all also central to being rural Creole. Facebook posts made by people living in Chicago or New York often show a big iron cooking pot full of beef meat on a fire hearth (outdoor kitchen) from their latest visit to Belize, attesting to the centrality of this knowledge and experience to being Belizean Creole (also see chapter 7).

Beyond Animals: Plantashe

Some villagers, mostly older men, "make plantashe," or engage in shifting cultivation, traditionally walking long hours to their farm, clearing and planting the land, and returning to reap months later (see figure 5.4). This is backbreaking work. As roads and vehicles have become more common, greater numbers of men drive to their plantashes, but still do the planting and harvesting by hand. The extent to which people in rural Creole Belize have done this has varied throughout history, depending on the other economic opportunities available.

Figure 5.4 Corn plantashe, Mr. Rudy Crawford and his son Luke, Crooked Tree, 1993. Photo by Melissa Johnson.

But making plantashe, growing food, is a rural Creole tradition with roots in the very first days of African and European presence in this part of the world. In the 1990s, when I started spending time in Belize, many households depended at least minimally on some homegrown foods, even if a neighbor or sibling were growing the food and sharing it with them. While planting is still certainly common, it is not as central today as it was then. In the twenty-first century, as in the past, plantashe grounds are most commonly located some distance away from village centers (where most people have their houses), because the rich soils that support good agriculture tend to flood and get muddy, and are not the ideal place to situate one's house. Most farmers plant corn, rice, ground food, or tuber foods such as cassava, cocoyams, and yams, and "suckers" such as bananas and plantains. Some also plant red kidney beans and assorted vegetables (squash, tomatoes, okra, sweet peppers, or greens like calaloo). This kind of agriculture requires clearing a new piece of land every year, usually cohune ridge (areas where dense stands of cohune palm trees grow), which entails a significant transformation of lands in the area. At the same time, the parcels of land that are no longer used grow back quickly with dense vegetation, becoming high bush in a few short years. Among variations of this typical plantashe scenario, some villagers have at times kept "coconut walks" or "lime walks"—special areas, sometimes quite large, devoted to a particular perennial crop, most typically fruit. Many people maintain small fruit orchards both in their yards and on their plantashe, and some keep small kitchen gardens of medicinal plants. Almost all people who spent their childhood in rural Belize can describe an experience going to a plantashe, when they were children. Their practical engagement with the land and the know-how they possess about growing food are important elements of what it means to be from this place.

Cashew

Some rural Creole communities are known for being home to cashew. Traditionally in these places, women have developed the knowledge and skill needed to process cashew seeds (or nuts), and the bodily involvement in creating cashew seed is a badge of pride for those who engage in this activity. Cashews grow abundantly in the sandy soils of a few rural Creole villages. Crooked Tree is one of these and villagers are proud of the cashew seed they produce, and that have brought the community fame. One evening in the early 1990s, I visited Miss Mildred, a grandmother raising three grandchildren in a house and yard down the road from where I stayed in Crooked Tree. As I walked to her house, the evening sun caught a billow of oily, acrid smoke rising up from the blackening cashew seeds in a small wood fire on the sandy ground of her yard. Cashew husks are thick and difficult to break, and they contain a noxious oil similar to the irritants in poison ivy. If the husks are charred in a fire, they break off easily.

Mildred sat on a small bench beside the fire and in front of a large mahogany bowl full of the thick blackened husks that surround the cashew seeds inside. One by one she broke off the charred husks and plucked out the tender seeds, dropping them into an old coffee can at her side, her fingers blackened and peeling from repeated contact with the poisonous oil in the husks. Once she "roasted and broke" the pile of seeds in front of her, she then lightly browned the nuts on a baking sheet in her oven. Three young grandchildren watched while she worked, their appetites awakened by the sweet smell of freshly baked johnny cakes (a type of biscuit) wafting from the small wooden kitchen in the center of the yard. Green parrots flew in noisy groups overhead, fat from feasting on the cashew fruit that grows so plentifully in the village in May. This is a common scene during cashew season (April and May), and women are proud of the hard work they put in to process these seeds, their blackened and peeling fingers a testimony to their strength and toughness, a symbol of how they are "haad" not "soff" (see chapter 3).

The centrality of cashew processing to being a rural Creole woman is illustrated by Miss Julia's comments to me in the 1990s. Miss Julia had a steady and good job as a cook at a sport fishing resort on the Belize River, and supplemented that income with rental properties in Belize City. She was a relatively well-off woman in the village. During cashew season, she left her cooking job early each day, much to her employer's chagrin. She lamented the fact that he could not understand why she would leave her job early for this arduous work. But this was part of being herself, of being rural Creole and from Crooked Tree: not to process cashews would deny part of her identity. Processing cashews surely generates additional income, but it also confers a sense of identity and pride in being a strong rural Creole woman.

Entanglements with Other Plant Forms

Being rural Creole emerges out of relationships with plant life in a range of other ways. A number of men engage in commercial logging of either large straight pine trees from the pine savannah as posts or a variety of hardwoods from broadleaf forests, such as mahogany, Santa Maria, My Lady, bullet tree, and sapodilla. Others, often young men who do not have the equipment to log large trees, cut fence posts from logwood. The logwood-cutting days from the 1700s echo in the work of young men today. Yet other men cut pimenta palm to make fish traps, and, if they are poor, to make houses. An ever-smaller number of individuals in some communities also produce charcoal by making coal beds in which oak wood is slowly burned in sand with pimenta leaves on top allowing sufficient oxygen to flow to form coal (see Benya 1977). The area where the bed itself is made is also cleared and dug out, the bed left to smolder for several weeks, and the sacks of coal then sold in nearby towns and cities.

Figure 5.5 Mr. Harry Cadle making coconut fat (oil), Crooked Tree, 1993. Photo by Melissa Johnson.

Knowing which leaves, fruits, flowers, bark, and other plant parts to gather, and gathering those for medicinal purposes is also central to being rural Creole. Many women, as well as both men and women known to be healers, maintain this knowledge.[14] For example, people make infusions of various herbs to ease "pressure" (high blood pressure), to "purge" (clean out the body), or to treat a wide range of other ailments from diabetes to stomach aches.

Other people do the hard work of producing coconut and cohune palm nut "fat" (oil) by hand. Some individuals make these oils regularly, others only occasionally (see figure 5.5). Coconut and cohune oil are highly valued products, and, like cashew seed, their production most typically falls within the domain of women's work. Handmade oils are typically used at home, or given as gifts to friends or family, less frequently they are sold to shops and markets in towns and cities.

Other people gather and sell fruits that grow on their land (either in their house yard, or on their plantashe land). These include mangoes, avocados, craboo, breadfruit, ginnep, plum, and dozens of other types of fruit that grow in central Belize. Some of these fruits require a bit of processing before they are sold, such as craboo, which must be preserved in bottles with water. Many women also make jams or preserves from the fruits that grow in their yard (or sometimes from fruit that grows in other people's yards) to sell either in village shops, at village events like the Crooked Tree Cashew Festival, at cricket games in any village, in Belize City on the street, or in tourist shops.

Residents are entangled with the natural environment in a variety of other ways. Gathering firewood used to be a much more important and widespread activity, when people depended exclusively on "fiyah haat," outdoor fire hearths used for cooking and baking (see figure 5.6). By the mid-1990s, most kitchens had gas stoves and only a handful of people gathered firewood on a regular basis. However, during the course of the year nearly every household still uses a fire hearth at some point. Fire hearths can accommodate the enormous pots that are used to cook rice and beans and to stew meat (from turtle to chicken, beef, pork, or deer) for large crowds. Frying fish on a fiyah haat also lets the odors of that mode of cooking waft away in the breeze. But it is also simply that food cooked on the fire hearth, like rice and beans, fried fish, or baked johnny cakes, tastes better to most Belizeans. Knowing how to make a fiyah haat, which kinds of wood burn best in it, and how to manage the heat for cooking are all essential knowledge and skill that rural Creole women and men hold.[15]

Firewood is also used for a range of other tasks, from boiling a large vat of water to singe the hairs off of a freshly shot peccary to roasting the husks off of cashew seeds. I have left out dozens of more-than-human engagements through which people become rural Creole, and I make this point because it is the very wide array of relationships with the more than human that characterize what it is to become rural Creole. But it is not merely that rural Creole people are

Figure 5.6 Fiyah haat in Crooked Tree, 1994. Photo by Melissa Johnson.

entangled with the more than human. The ways in which these entanglements assemble with social relationships and cultural productions also generate rural Creole culture.

Economies and Ecologies Otherwise

Rural Creole everyday life is intimately entangled with the more than human. But rural Creole socionatural engagement reveals other dimensions of what it can mean to be human in the world; dimensions that are often underreported, unseen, illegible, and undernoted in much scholarship on the Global South in this neoliberal moment. A close look at this part of Belize, a place that has been embroiled in capitalist economic organization for the past five hundred years, as I show in chapter 2, reveals that while market-based production within regimes of capitalism has always been a feature of rural Creole life, it has never been *the* defining feature of that life. Rural Creole Belizeans' economic life is also organized through noncapitalist forms and practices. And this is especially the case in economic activity centered on the more than human. It is in the exchanges of elements of the natural world that noncapitalist relations are most prominent in rural Belize.[16]

As elsewhere in the Caribbean, a sense of community and connectedness does not primarily emerge from a shared sense of corporateness, as might be the case for the Maya in Belize, for instance, which unites people through a sense of being one body of people tied together deeply through shared rituals, symbols, myths, and clearly defined kinship lineages. The Belizean Creole sense of community develops instead from a rich and overlapping network of exchange and kinship relations (see Rubenstein 1987). Exchanges among kin, neighbors, and fellow villagers in general are central aspects of the social life in rural Creole Belize, creating community and generating senses of belonging.

On Sharing

Sharing and an ethic of generosity are organizing values of rural Creole culture. This is well illustrated by a scene I have witnessed many times over the years in many different settings, and that I first found extraordinary—and challenging—to witness. Babies are taught from a very young age that being greedy or selfish is not socially acceptable. A baby will be offered a treat—a piece of mango, or a piece of candy or cake—and then someone will say to the baby: "Gimme some" or "Please fi some." If the baby does not offer the treat back, the person takes the treat away, and if this is early in the baby's training, the baby howls and screams. The person may or may not take a bite, and then gives the treat back. After several such experiences, the baby becomes calm and is more than willing to share the special treat. These scenarios of a baby's first experiences with this situation jarred me the first few times I witnessed them. But I began to appreciate them as I learned that this was one way in which the rural Creole ethic of generosity that I admired so much was created. My husband and I subsequently tried using this Creole baby-rearing technique in raising our own Belizean American children. Yet I also know I often do not fully inhabit this ethic myself, and my family members have to remind me to be as fully generous as a rural Creole person.

This ethic of sharing is perhaps most fully elaborated in the exchange, generalized reciprocity, and gifting that mark engagement with the more than human in rural Belize. Freshly hunted game meat, freshly caught fish, fruit that has ripened in a yard, or foods made at home from local products are prime items for trade, material items that, as they are shared and exchanged, bring people together. While some individuals hunt and fish expressly to sell the game and fish to stores or wholesalers, most hunt and fish less intensely and share their catch, or sell it in person to someone they know well or are related to in the village. Even hunters and fishers who sell commercially also share or sell locally. Exchanges of game meat and fish are still often generalized, there is no expectation of timely return of a similar value. Fish or meat from a hunt is shared first with close kin, neighbors, and friends, and then, if any is left, with more distant social relations in the village.

But sharing the spoils of the hunt is not unfraught. Friends and family watch very carefully to see who gets how much, and evaluations of the appropriateness of the gifting of game meats can be debated long after the meat was shared. For example, in January 2015, in Texas my husband and his cousin in New York were having an animated telephone conservation about how a peccary had been shared in Lemonal the previous week. The hunter did not share half the peccary he caught with the two men who accompanied him on the trip, and instead gave it all to just one relative, an inappropriate way to share the meat, according to Elrick and his cousin. Rural Creole people are very careful in figuring to whom they give what, but gibnut, deer, peccary, hicatee, and choice fish are particularly subject to this close evaluation. This care and attention underscores the importance of the exchange of the more than human to maintaining the social ties that create Creole culture in rural Belize.

Memories of large-scale exchanges of these sorts and of hunting parties working together peppered my interviews with people in rural Belize in the mid-1990s. One schoolteacher lamented the loss of the greater sense of community that she remembered existing in the past:

> Tommy, Ray and wahn man fi Uncle Herbert used to go hunting together. People from round mi a come and get fi they cut. They mi bring tortilla an mek wa big pot a black tea. My ma kitchen mi a full. The ladies come again in the evening fi mek wa big pot a soup. Ih wa be like wa ritual, ih no hapn every day. . . .
>
> Anytime anyone mi have wahn special ting fi cook, wahn piece a meat or so, no matter how small the piece was, a plate with a little bit on it would be sent around to neighboring houses.[17]

> (Tommy, Ray, and one of Uncle Herbert's sons used to go hunting together. People from this part of the village would come and get their share of meat. They would bring tortilla and make a big pot of black tea. My mother's kitchen would be full. The ladies came again in the evening to make a big pot of soup. It would be like a ritual, it wouldn't happen every day. . . .
>
> Anytime anyone would have something special to cook, a piece of meat or something like that, no matter how small the piece was, a plate with a little bit on it would be sent around to neighboring houses.)

These events of sharing and togetherness centered on wild game meat are some of the fondest memories that rural Creole people hold and are central to their sense of themselves and their communities. While exchanges like this still occur, they are less common than they were, as more money flows through the community from migrants' remittances, and migration to the United States changes and attenuates these connections.

Even when people do not share, but rather buy and sell meat and fish, both buyers and sellers appreciate the possibility of personal exchange with their neighbors rather than buying from a store. In recent years, with the advent of electricity in rural Creole Belize, shops sell meat, but villagers still prefer fresh game and fish to buying meat from the shop. People appreciate variety in their foods, and game meat and fish also tie them to their rural roots. The importance of this type of exchange is revealed most clearly when it does *not* happen, when people fish or hunt and then do not sell to their neighbors, or when there is little fish or game available. In these instances huge cries of complaint are heard: "Weh wi wa cook? How Paulie figot wi?" (What will we be able to cook? How could Paulie have forgotten us?). And even though more-than-human nature may be exchanged for money in rural Belize, these exchanges are still intensely personal and social and serve to maintain social relationships. The hunter tells the story of how he caught the peccary even as he takes the dollars his neighbor gives him to pay for the meat. The person seeking fish sees the fisherman paddling in and chats with him as his child loads the bucket to sell fish door to door.

Wild game and fish dominate these networks of exchange, their sudden emergence delightfully unexpected and serendipitous, the hunter and fisher never certain they will meet success in their endeavor. But other forms of more-than-human nature also circulate among rural Creole Belizeans. Beef meat serves as a gift, as social glue that ties people together. The exchange of beef meat is typically more planned and arranged, sometimes years in advance (a steer offered for a wedding, or a christening, for example). Cattle and beef meat have a long historical presence in this part of the world, as described in chapter 2. As noted above, cattle serve as a bank account for many rural Belizeans (and indeed for many people all over the world), repositories of value, and also indicators to others of wealth. Cattlemen tend to members of the wealthier households in most Creole communities, and cattle insert rural Creole people into capitalist economic relations. They can be used by the entrepreneur who wishes to expand holdings, re-investing profits in expanding the firm—the ranch. But at the very same time, cattle are a critical part of economic activity *otherwise* and the very important gift economy of rural Belize. Beef meat is highly prized and is not everyday fare like chicken. When someone plans a wedding, wake, or funeral, beef is the preferred meat to serve. It is not uncommon for a close relative or friend to donate a young bull for an event such as this. Sometimes a bull will be sold rather than freely given, but usually at a low price, and to a fellow community member. The structures of feeling regarding the relationship between the cattleman and the friends and family receiving the meat are similar whether the meat is bought or shared. Cattle are also given alive, for example, when a cattleman gives a heifer to a young relative as an economic foundation on which to build. Other animals are given as gifts as well (horses, sheep, goats, and pigs, for example). In all these examples, the exchange of elements of the more than human serve to create more and closer ties between people.

Exchanging Labor and the More than Human

Working together to fish or hunt, clean land, catch cattle, and so on is critical to how kin and friendship ties are maintained, and constitutes another way in which noncapitalist economic relations organize life in rural Belize. When individuals hunt and fish and they are accompanied by friends and relatives who help them out, those friends and relatives typically share either in the money earned if the fish or meat is sold, or they take home some of the catch. Nearly every man volunteers to help kin, neighbors, or other male friends in work with the more than human, helping to chop plantashe, clean a peccary, or haul a large fish net, for example. Working cattle is difficult to do individually, and men often help each other catch young calves, build cattle pens, and drive cattle into new pastures.

One common occurrence is the all day, or several days, work group. A group of eight to fifteen men all contribute labor to a substantial task—such as fencing a big yard, or building a large cattle pen or a small house. Some men work hard for a stretch, while others watch on, then they shift, but all men do not necessarily work equally hard—contributing to the sociality and fun is also important. Everyone involved is chatting, often drinking rum, telling jokes, and laughing. Women and children gather around at times to watch and enjoy the sociality. The women in the household of whoever has organized the work cook a large amount of food for all to eat. This "work" then is as much socializing and fun as it is work. Whoever benefits from their friends' and relatives' labor is expected to give back to the others when they need help. But there is no keeping tally and no expectation of immediate return. People note who are the more "willing" (willing to provide their labor) of the men, and those men are appreciated for being willing, but men who give less of their time and energy are not criticized for it.

Women I interviewed in Crooked Tree talked about how sweet the days in the past were when groups of women would go together to one of the ponds in the village to wash clothes—spending a good part of the day together, outside of their houses, washing clothes, talking and visiting with one another. What would be onerous work otherwise, then became social fun.

Sometimes, shared labor with the more than human is smaller scale and routine. On many evenings in the early 1990s in Crooked Tree, I would join the family of Mr. Raymond and Miss Linda, picking (or shucking) red kidney beans outside the house in the cool shade, while people "talked fool," making clever jokes and word puns while they worked. If someone came to visit they might join in on the work, but they most certainly would join in the joking and storytelling. Were these hours "work" or "fun"? The lines between the two are blurred.

Entanglement with the more than human, sharing both the fruits of these entanglements and the labor they necessitate, generates a sense of what it is to be rural Creole. But exactly *what* things emerge from the engagement of person and ecology in this part of Belize also generates a *particular* sense of community identity.

Community Identity

Interestingly, people have a sense of what it is to be from rural Creole Belize even if they do not spend much time engaging in the activities I describe above. Rural Belize is similar in a number of ways to fishing communities in Donegal, Ireland, as described by Lawrence Taylor (1981).[18] Even though a typical man in Teelin, the village Taylor described, was likely to be a wage worker in a factory, or a small farmer, who in a full year may spend only eight or ten hours on the water fishing, if he was asked what it was he "does," he would reply "I'm a fisherman." The village was known in Gaelic as "teelin of the fish." Similarly, while many rural Creole people work outside of rural villages as wage laborers, civil servants, or may even live elsewhere (places as far-flung as Belmopan, Phoenix, or Fiji), they all spend at least a little time engaged in the kinds of relationships and activities I describe above and these activities loom large in their thinking. For those who have left Belize, fishing in parks outside of Chicago or processing meat or turtle brought to them once a year from Belize by relatives may constitute the bulk of their embodied engagement in a rural Creole way with the more than human (see chapter 7 for an elaboration of these points). Those living in Belize, even if they spend very little time fishing, hunting, and tending cattle, are still connected to the more than human of these rural areas through the work activities of family and friends who live there.

This sense of what constitutes rural Creole identity struck me when I asked people in Crooked Tree about their thoughts regarding conservation over the years. Many responded adamantly that first and foremost they were fishermen and hunters, and said things like: "Killing deer in Crooked Tree is like the running of the bulls in Spain," "Its village life, if yu pa kill animals, and you tek away that, that no de wahn village again" (It's village life, if your father killed animals and you take that away, then it's not a village again). These statements implied that limiting the ways in which people use the natural environment, in particular in their hunting and fishing, would destroy the community and Crooked Tree residents' sense of what it is to be a rural Creole person from this village. This sentiment is shared throughout the communities in this area. And, as I describe in more detail in chapter 7, diasporic rural Creole Belizeans continuously reaffirm their rural Creoleness by celebrating fishing and hunting through social media, telephone and video-chat conversations, and gatherings in various locales around the United States that often feature fish and game meat from rural Belize.

Particular fish, meat, or foods are often associated with specific villages, so that a sense of belonging to a particular village is tied to a given element of the more than human. If you mention to someone in another village that you are going to spend the weekend in Crooked Tree, they will inevitably ask "and go drink some of that cashew wine?" and "eat a lot of fried fish?" As noted above, Crooked Tree is associated with crana and bay snook, the fish commonly found in the lagoons around the village; tilapia have joined that list in recent years. Crooked Tree is also famed for its cashew seed, cashew wine, and other cashew products. If you visit Lemonal, people will ask whether you ate deer or drank the black water of Spanish Creek. Knowing how to make steamed flour over stewed local chicken also marks you as someone from Lemonal. Rural Belize is home to a particular kind of bushy village life that is shared across all communities, but each community is also known for specific elements of the more than human.

Above I have described just a few of the ways in which rural Creole people become enmeshed with the more than human, and how in these ways they become people who are rural Creole. But it is not only through physically engaging that people bring rural Creole into being: the more than human also occupies the imagination of rural Creole people.

Language, the More than Human, and Becoming Creole

The embodied engagements and becomings with the more than human that I describe above are re-inscribed through everyday storytelling and linguistic performance. Rural Creole Belize shares with other Afro-Caribbean cultures a celebration of prowess in mastery of the spoken word in everyday life (Abrahams 1983, Cooper 1995). Men, in particular, compete in casual and friendly ways around a bar table or in chairs in a yard to hold the attention of a crowd by being the most engaging and spirited storyteller, speaking the Belizean Kriol language, and using gestures and sound effects to pull the listener in. Very often in the communities of rural Creole Belize, the story being told is about hunting, fishing, or a trip in the bush. The storyteller uses the rhythm of language, speaking loudly, mimicking the "pow" of a rifle going off, the deep grunt and growl of a halligata, and making liberal use of the metaphor, simile, and double meanings that mark the Belizean Kriol language. At the club, where men congregate, as a bottle of rum is passed around and the bass line of dancehall pounds through the room, discussions turn to stories of hunting, fishing, cattle-rearing, and planting as often as to anything else, and everyone has a tale to outmatch the speaker before. Men build their reputations not only through their abilities to hunt and fish or their knowledge of the backwaters and bush surrounding the village, but also through their ability to tell stories about their exploits with the more than human.

The natural world has rich metaphorical meaning for rural Creole Belizeans and it figures centrally into the double entendres and ambiguities that provide a source of both humor and social critique in Creole languages (Abrahams 1983; Cooper 1995; McAllister 2002; Smith 2001). Given the myriad ways in which the more than human is meaningful and important to the people from this place, it should not be surprising that various plants, animals, and natural features serve as key metaphors and sources of social meaning for rural Creole Belizeans and that their language play depends deeply on these metaphors. Even a mocking Kriol name for the Kriol language uses such a metaphor. "Bocatora language" is what Creole people derogatorily but playfully call the version of Kriol that is farthest from standard English.[19] The bocatora is the slowest and easiest to catch of the land turtles, reminding the speakers and audience of the elite framing of Kriol as being backwaad and bushy. Looking for resemblances between people and animals is a favorite and amusing pastime for many rural Belizeans. The resemblances are based both on the physical features of a type of animal and on its behavior. For instance, an irritable person is described as a peccary "knocking" its teeth in warning and miserableness. "She da wa direck mountain cow" (She looks exactly like a tapir) describes a fat woman with "no shape," or lacking the small waist and large bottom and thighs so admired in the Caribbean. "Ih no fava wa bocatora?" (Doesn't he look like a turtle?), a friend asks as they laugh at the looks of someone else. "Ih like play lone quash" (He acts like the solo coatimundi) describes the antisocial behavior of a young man rumored to have many women, by referring to the single male coatimundi who has not yet found a social group of which to be a part. Children call each other "maka parrat" when one mimics another, referring to the macaw parrot, but macaw is a homonym with the Kriol word "mocka," or mocker, meaning someone who mocks. Commenting on the 2016 NBA (National Basketball Association) finals, my husband laughed and loudly declared to my sons that Lebron James of the Cavaliers would not be able to win the championship: "Ih no gwine noweh, ih head too peel, fava pelican" (He isn't going anywhere, his head is too bald, looks like a pelican), and then later, "It's not the horse that takes the lead, it's the torobred from behind" (Its not the horse ahead at the beginning that will win, but the thoroughbred who comes from behind). These clippings of everyday speech show how the more-than-human world figures deeply into the everyday thinking of rural Creole people. Linguistic play with animal and plant names can be among the most witty and amusing, as well as among the sharpest social and moral critique. In this way, the more than human serves as a kind of touchstone for the people of this place, a measure against which things can be judged and made meaningful, ironic, and amusing.

Music

The more than human figures not only into language play but is also very much a part of Belizean musical traditions. One of the most popular folk songs that all Belizeans learn at a young age is about how challenging the "bush" is to navigate because of the power and agency of the plants and animals it contains:

Ping wing juke me, mosquito bite me,
cyah go deh at all,
Wah go to mi plantashe,
cyan go deh at all . . .

(Ping wing [*Bromelia penguin*—a sharp spiny low plant] pricks me,
mosquito bites me
Can't go there at all
Want to go to my farm
Can't go there at all . . .)

Contemporary musical production also often incorporates the more than human. One of the leading figures in Belizean "brukdown" music, the music most closely associated with Creole culture, Mr. Wilfred Peters (and his Boom and Chime Band), released a song in 2006 celebrating Crooked Tree and its socionature.[20]

The lyrics:

Crooked Tree bear di biggest cashew
Crooked Tree catch the most bay snook
Crooked Tree catch the most tilapia

I could mention numerous other songs, but one in particular reflects a number of themes of this book. The band Caribbean Dynamics formed by a group of Creole, Garifuna, and Mestizo/Spanish Belizeans in Los Angeles in 1997, released a song in the mid-2000s entitled "Mr. Banner" about the particular problem of jaguars plaguing rural Creole villages (one of the band members was/is a Banner from Camalote Village in Belize's Cayo District):[21]

This is a true true story
Way down in Lemonal
There was a tigah in the area
Killing Mr. Banner calves
So ih went up to Cayo [the westernmost district of the country] and get
ihself a tigah dog
Then back to Lemonal and ih hunt di tigah with the dog
Mr Banner, Mr Banner, Mr Banner, tigah gone with your dog . . .
Way down in Lemonal, tigah gone with your dog

Run [Risdale] Run, tigah gone with your dog
Run [Frankie] Run, tigah gone with your dog
Run [Turo] Run, tigah gone with your dog
Run [Bannah] Run, tigah gone with your dog
Run [Primo] Run, tigah gone with your dog
Run [Marlon] Run, tigah gone with your dog

So Mr. Banner got fed up and figet to hunt fi tigah
Searching through the area all he found was dead dog
So from that day they call him
Mister Banna Mister Banna Mister Banna, Tigah gone with your dog

Way down in Camalote
Killing Risdin little dog

These lyrics are predicated on both Lemonal and Camalote being more rural, more out-of-the-way, than other places; they are places where jaguars are feared and must be hunted down, but only after one has gone to the city (Cayo) to get a good hunting dog.

These are just a few examples to illustrate the general point: the more than human serves as a critical source for cultural production, for providing meaning to rural Creole people. From competitions to hold an audience's attention through storytelling, to joking and punning, and woven throughout different musical forms, the more than human forms the stuff of creative and artistic production for people who identify with this part of Belize.

Conclusion

I have described in detail how the people who live in rural Belize become who they are through their engagement with the more than human. People from rural Belize have great pride in their abilities, knowledge, and competence among the more than human, and derive their sense of who they are from this enskilment. I argue that while Belizeans may produce commodities from the more than human for distant markets, they also exchange fish for deer with a neighbor, for example. And it is these noncapitalist exchanges that create meaning and community for rural Creole people. Particular elements of the more than human, for example, Spanish Creek's black water and Crooked Tree's cashews, mark certain places and certain communities. Finally, the more than human serves as the backdrop for a rich set of meanings and metaphors through which human foibles and strengths can be understood, and rural Creole culture can be celebrated through story and music.

Rural Belizeans exemplify ways of living *otherwise*: people become rural Creole Belizean by enacting an ecological personhood, one that comes into being relationally with the more than human. This mode of being human in the world cannot be fully comprehended through the assumption of the universal human subject that is still understood Eurocentrically: the autonomous, rational, self-serving independent being fully separated from the world around it. I suggest that the way of being human that is so apparent in rural Belize constitutes one of the multiple genres of human being to which Sylvia Wynter refers (2003). Recognizing this mode of being, making it legible, moves humanity collectively from privileging relationships of dominion over the more than human to more equal relationships with the world beyond the human.

One other dimension of the becomings with the more than human that bring rural Creole people into being is their embodiment. The repeated assembling of Creole people who have brown skin of various shades, a range of facial features and hair types, with the deep knowledge, skills, and sociality necessary to live well in this place generates associations of brownness and blackness with positive values.

CHAPTER 6

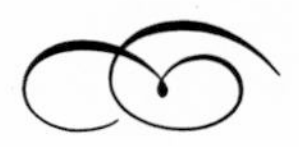

Wildlife Conservation, Nature Tourism, and Creole Becomings

In June 2016, a YouTube video was posted on Facebook by a dear friend from Crooked Tree who now lives in Chicago. It was a Belize news report from the day before. Crooked Tree fishermen were stopped from hauling their fishing nets close to the causeway into the village, even though they thought hauling there was legal. Earlier in the day, I had heard something might have happened after my father-in-law had asked my husband on the phone if he had heard about the problem with fishing in Crooked Tree, and so I watched with great interest. In the video, the reporter, a large dark-skinned Creole man, interviewed a few young Crooked Tree fishermen. A range of other people, men, women, and children milled about in the background, including the typical mix of skin colors and embodiments of this place. Farther in the background were police officers, a sanctuary warden and other government representatives in their official attire, leaning on a well-kept shiny new pickup truck. A young dark-brown–skinned fisherman told the reporter:

> This affect we big time because we as a ghetto youth, this dah fi we living weh we do from tradition; this dah pa, ma, everybody di do this. And I noh know weh di go on with dehn bally dehn. But Audubon dah somebody weh we need fi take out yah because dehn di try stop fi we livity. And if dehn di try stop fi we livity, weh dehn want we do? Dehn want we go bruk ina people house? . . . When we go look fi wah job, first thing yo hear is that we dah criminal with the way how we look; di try class we bad. (Channel 5 Belize News—their orthography of Belize Kriol)

> (This affects us big time, because, us, as poor black youth, this is our living, what we do from tradition, this is father, mother, everyone was doing this. And I don't know what is going on with those guys [the police/wardens]. But Audubon is somebody that we need to take out of here because they are

> trying to stop our livity [livelihood, way of life, with essence of Jah, living in love and nature, full being-ness]. And if they try to stop our livity, then what do they want us to do? Do they want us to go break into people's houses? . . . When we go look for a job, the first thing you hear is that we are criminals with the way we look, they try to cast us as very low class.)

This fisherman expresses his rural Creole engagement with the more than human, his "livity" and the efforts of conservationists to limit that engagement.[1] He also smartly notes how people from rural areas are seen by city dwellers, a point that resonated powerfully with my husband as we watched the video together. The film then cuts to the office of the Belize Audubon Society in Belize City and its director—a light-skinned Mestizo woman in tidy office attire. She at once defends the actions of the government and Audubon while also claiming that Audubon intends to work with Crooked Tree fishermen and women to allow them to fish in lagoon waters.

One warm day in the mid-1990s, after Crooked Tree had been home to a wildlife sanctuary for about ten years and had become established as a site for ecotourism, I walked out onto a dock over the rippling waters of Crooked Tree Lagoon to feel a little breeze and enjoy being close to the water. A light-skinned teenager from the village was fishing from the dock. He had caught a few fish—crana and tuba—and had left his catch on the dock, flopping as they slowly died. He was concentrated on fishing and was not worried about the fish. An elderly couple—tourists, white, speaking with American accents, walked out onto the dock at the same time. They had binoculars around their necks and were eagerly exploring the place they would be spending the night. As they spotted the fish flopping, they expressed their discomfort, even disgust, at seeing the struggling fish. Just within hearing distance, the teenager had witnessed the tourists' reaction, and as they turned, he chuckled in a self-satisfied way and threw another fish onto the dock. The news video and this anecdote reveal the connections between Belizean entanglements with the more than human, processes of racialization, and the biodiversity conservation and nature tourism that have become part of everyday life in rural Belize.

In the preceding chapters I have shown the historical development of rural Creole entanglements with the more than human, how they have generated economies and ecologies otherwise, and how these entanglements are racialized. In this chapter, I examine how the global phenomena of conservation, on the one hand, and tourism on the other, assemble with socionatural entanglements and processes of racialization in Belize. Encounters with conservation and tourism inflect what it means to be rural Creole today at the same time that rural Creole people continue to create economies and ecologies that are not overdetermined by capitalist logics. These ecologies and economies assemble with rural Creole people's range of embodiments in ways that sometimes re-enforce and other times break down dominant racial orders.

Global momentum to protect nature coincided with the popularization of tourism in the mid-twentieth century, and Belize was caught in these currents from the beginning (Boardmann 1991). The lagoons and waterways of Crooked Tree and Spanish Creek were some of the first places in Belize where these currents converged. This part of Belize, characterized by seasonally flooded lowland savannahs, endless rivers, and enormous flocks of every kind of waterfowl imaginable, including the six-foot-tall jabiru stork, was an early attraction for U.S.-origined middle- and upper-class white ecologists and birders, as well as local elites who partook in the growing popularity of birdwatching as a leisure activity. This interest contributed to the creation of a wildlife sanctuary and the development of an ecotourism industry in the area by the early 1980s, and the consequent interweaving of conservation and tourism into rural Creole Belizean lifeways (Johnson 2015).

Conservation, and in particular the setting aside of protected areas, has been well studied in the scholarly literature (Brockington 2002; Doane 2012; Grandia 2012; West 2006; West, Igoe, and Brockington 2006). Researchers describe how biodiversity conservation projects disenfranchise local populations, often reproducing colonial relations of power. International conservation organizations with headquarters in the United States and Europe typically work with governments of countries located in the tropics to implement conservation plans. People living in targeted areas are sometimes removed from ancestral lands as parks are put into place, or residents are displaced by virtue of no longer being allowed to hunt, fish, or collect natural products from protected areas. In the most egregious situations, military and police forces have been used to keep people from their lands, creating what Dan Brockington has called "fortress conservation" (Brockington 2002). In seemingly more benign efforts to set aside land for biodiversity conservation, local populations are encouraged to take part in new economic activities (tourism or other income-generating projects), or to co-manage their lands with government entities. But even in these situations, the logic of the market, privatization, and capitalism can transform local economies and the people who live in them: neoliberal policies can create new worlds of alienation and individuation (e.g., Doane 2014; Igoe and Brockington 2007). In this chapter, I ask a slightly different question about the presence of conservation and tourism, inspired by Gibson-Graham's exhortation that scholars should think beyond and around capitalism to see other forms of economic activity (Gibson-Graham 2011, 2008) and Sylvia Wynter's critiques of Man (McKittrick 2006, 123; Wynter 2003): What else is going on? Do people also maintain their socionatural becomings with the more than human when conservation enters the picture? How do they weave conservation initiatives into their everyday entanglement with the more than human? Do they seek out, reach out, toward conservation and tourism possibilities (see Hathaway 2013)? Do they continue re-creating long-historied socio-ecologies alongside conservation? I also explore the relatively understudied issue of how conservation initiatives are racialized (but see Sundberg 2008a, 2008b; Suzuki 2016). How do racial formations shape conservation in practice and how do conservation encounters assemble racial formations? In Belize, conservation has been linked to nature tourism from the beginning, and I ask similar questions about ecotourism. Scholars analyzing ecotourism have suggested that this industry commodifies nature, and contributes to the alienation of people from nature (Bandy 1996; Davidov 2013; Duffy 2013; Fletcher 2014; West and Carrier 2004). How does ecotourism assemble with rural Creole socionatural becomings and, relatedly, how do nature tourism encounters highlight what it means to be rural Creole? In Belize, ecotourism, like conservation, is always a racialized process, and here I show how it assembles with blackness, brownness, and whiteness.

Conservation and Rural Creole Socionature

The setting aside of protected areas in Belize is built on a much longer history of conservation, in which the lower Belize River Valley plays a central role. As is the case elsewhere in the Global South, conservation initiatives, whether coordinated by the Colonial Office, the government of Belize, or eager nongovernmental organizations (NGOs), have always been tightly tied to capitalist forms of extraction and appropriation of the natural, whether to ensure the steady flow of timber products or to attract bird-watchers. By the early twentieth century, the historical record reveals a growing concern from colonial officers about the conditions of the colony's natural resources, especially Belize's forests and mahogany (Hartshorn et al. 1984; Platt 1998; Wright 1959).[2] Deer hunting also garnered interest early in the twentieth century (Munro 1983), and centered at least partially on the lowlands in the northern part of the lower Belize River Valley. Colonial officials noted what they saw as overhunting of deer using dogs and headlights along Spanish Creek and the lagoons of Crooked Tree. Legislation was enacted in the 1920s, and rural Creole Belizeans were subject to being stopped by police and colonial officials. In the 1950s, Randolph Tillett, a scion of one of the founding families of Crooked Tree, represented villagers' concerns, namely, their rejection of colonial regulations that limited their capacity to hunt. Tillett became a beloved historical figure in Crooked Tree, in no small part because he managed to secure a respite from limits to deer hunting for villagers. This early encounter already cast conservation as at least partly about the wishes of outsiders, elite and often white, or lighter-skinned colonial officials, impinging on the activities of rural Creole people, and highlights how rural Belizeans continued to assert their own way of being in this landscape, as their ancestors had during the early days of logwood extraction.

By the mid-twentieth century, the nascent international conservation movement joined with the growing importance of tourism in planning for national parks in Belize (Munro 1983, 143). Today, 26 percent of Belize's land and sea territory is protected in some way. Not coincidentally, Belize continues to have relatively healthy populations of many species that are otherwise endangered in the Americas.[3] The first areas targeted for conservation were the lagoons and waterways that are the focus of this book. These lagoons had been visited by U.S.-based biologists and bird-enthusiasts who were especially excited about the magnificent and unusual jabiru stork. Entrepreneurial villagers from Crooked Tree made the most of this early interest—selling transportation services to the nature enthusiasts, weaving guiding and transportation into the suite of activities from which they made livings, and using their knowledge of the landscape to bring in these outside experts. From the beginning, residents of Crooked Tree made particular note of two aspects of the relationship they began to develop with U.S.-based scientists and conservationists. First, rural Creole people were very aware of the whiteness of the U.S.-based scientists, always specifying their race when describing them (white lady, white man, white people). The descriptor "white" racialized these people as not Creole, from "foreign" and undoubtedly resonated with the long history of slavery and British colonialism in Belize. Second, they experienced these white scientists as "know-wells" (know-it-alls). The U.S. scientist most central to establishing the sanctuary was remembered by one older Crooked Tree resident: "She was all right, but she was a know-well, too." U.S.-based scientists and conservationists have typically operated under the assumption that they are the experts who know more about these birds and this ecology than do the Creole people living here.

During the early deliberations that led to the establishment of the sanctuary at Crooked Tree in the late 1970s, residents made very clear to conservationists and ecologists that they would not agree to giving up their fishing rights. This point was duly noted, underscored, *and* highlighted, by one of the U.S.-based white consultants working on the sanctuary's creation (Deshler 1978).

> It is recommended that the area should be established as a National Reserve to be managed as a bird sanctuary with provision to permit THE CONTINUANCE OF THE ESTABLISHED COMMERCIAL FISHING PRACTICE. (Deshler 1978)

While villagers may have had concerns about establishing the sanctuary, they also saw opportunity. They thought it might speed up a long-desired connection to the growing road and highway system of the newly independent nation, the village having been accessible only by boat prior to the 1980s. Crooked Tree residents are on record as having petitioned for road access from the 1950s (Minute Paper 1016 of 1951). Residents argued that theirs was one of the most important farming villages in Belize, and thus deserving of a road. The Northern Highway was rerouted in 1979 much closer to Crooked Tree, bringing the possibility of road access tantalizingly close. In 1984, the same year the sanctuary was legally declared, a causeway was built across the lagoon, bringing residents of Crooked Tree within an easy one-hour drive of Belize City.

The possibility of road access was not the only appeal a newly declared wildlife sanctuary offered. The sanctuary would also be a celebrated part of Belize's entry into a tourism economy, simultaneously highlighting the forward thinking of villagers and connecting to their sense of being central to Belize's national identity by virtue of their location as one of the first woodcutting areas in the colony.[4] Thus conservation was a means by which people in Crooked Tree could both live a more cosmopolitan life and reclaim their community as a place central to the nation. These simultaneous local and transnational interests reflect how rural Creole socionatural engagements have always entailed simultaneously reaching out to the possibilities offered by global economic interest and retaining the alternative ways of relating to the more than human that their ancestors had developed in these watery lowlands.[5]

Conservation initiatives also necessarily engaged processes of racialization from the outset. Historically, conservation and the setting aside of protected areas have been U.S.- and European-centered projects. The idea of setting aside protected areas was championed by the likes of John Muir, whose underlying racism and Eurocentric orientation has recently been brought to light. Indigenous peoples and people of color were seen as destructive interlopers in these natural places that needed protection (Deluca and Demo 2001; Finney 2014b; Merchant 2003; Sundberg 2008a, 2008b). As conservation became transnational, these same imperatives foregrounded interests of white, European-ancestried people and erased the legitimacy of claims on land and resources made by nonwhites. This was as true of Belize as anywhere else. All the earliest deliberations about conservation were vocalized at least partially by white British colonial officers or visiting white American scientists. The tendency for whiteness and light-skinnedness to assemble with conservation initiatives continues today. Governmental and nongovernmental positions in conservation are steady jobs that pay well in a country that has chronic high unemployment. The color-class hierarchy in Belize ensures that lighter-skinned people (whether light-skinned Creole, Mestizo, or white expatriates) are disproportionately represented in these (and any well-paying and steady) jobs.

Yet while conservation may have served as a vehicle to shore up racial difference and solidify white dominance, it did not work quite so simply as this in Crooked Tree. The history of the establishment of the sanctuary at Crooked Tree also served as a way for rural Creole people to racially spatialize the lands around the sanctuary, to limit some of the more distressing aspects of whiteness that they had been experiencing in the mid-twentieth century. In the late 1970s, just prior to the establishment of the sanctuary, white male sport hunters from the United States, Texas specifically, who were staying in Belize for an extended time, came regularly to Crooked Tree to hunt waterfowl. They often chose to hunt along the edge of the lagoon in the center of the village. Villagers were not pleased about this. Not only did these Texan hunters scare away a popular game bird, the whistling duck, but villagers were also unhappy because these white men often drank heavily when they were hunting birds. The story goes that one time the local Creole police constable went to the lagoon side to try to talk to the hunters to make them stop their drunken shooting. The hunters responded by turning around and aiming their shotguns at the constable! These white men had a sense of entitlement to this place—they felt they belonged here and could do as they wanted, unfettered. They dismissed the black residents, most likely not considering them as equal. Villagers were indignant at this treatment. The encounter between the constable and the hunters rallied villagers to support the creation of the sanctuary: it would provide an effective means of keeping out unwelcome outsiders, limiting what male bodies with thin noses, lips, and hair and very little melanin in their skin, most typically coming from "foreign," could do in these spaces. It would also limit the ability of outsiders of any and all types to exploit the natural resources, birdlife and other wildlife of this place.

With the establishment of the sanctuary, it became illegal to hunt, fish, or in other ways procure wildlife (animals, birds, reptiles, etc.) and plants in the lagoon and creek waters, and three hundred feet inland from the shore, with the agreement that villagers could continue their fishing practices.[6] The Belize Audubon Society, a Belizean conservation NGO that was created in the 1970s, was deputized by the government of Belize to manage the sanctuary (Waight and Lamb 1999). Two villagers were hired to work as wardens, a house along the lagoon edge served as sanctuary headquarters (see figure 6.1), educational activities were implemented in the village school, and Crooked Tree became a noted stop on birders' trails and in the growing nature tourism industry.[7] At nearly the same time, but through a less participatory process, a huge tract of land to the west of Crooked Tree and northwest of Lemonal, the Rio Bravo Conservation and Management Area (RBCMA) was set aside and protected, and is now managed by the conservation NGO, Programme for Belize (see map 6.1). Hunting and fishing were limited in the RBCMA, but like the Crooked Tree Wildlife

Figure 6.1 Entrance to Crooked Tree Wildlife Sanctuary, 2005. Photo by Melissa Johnson.

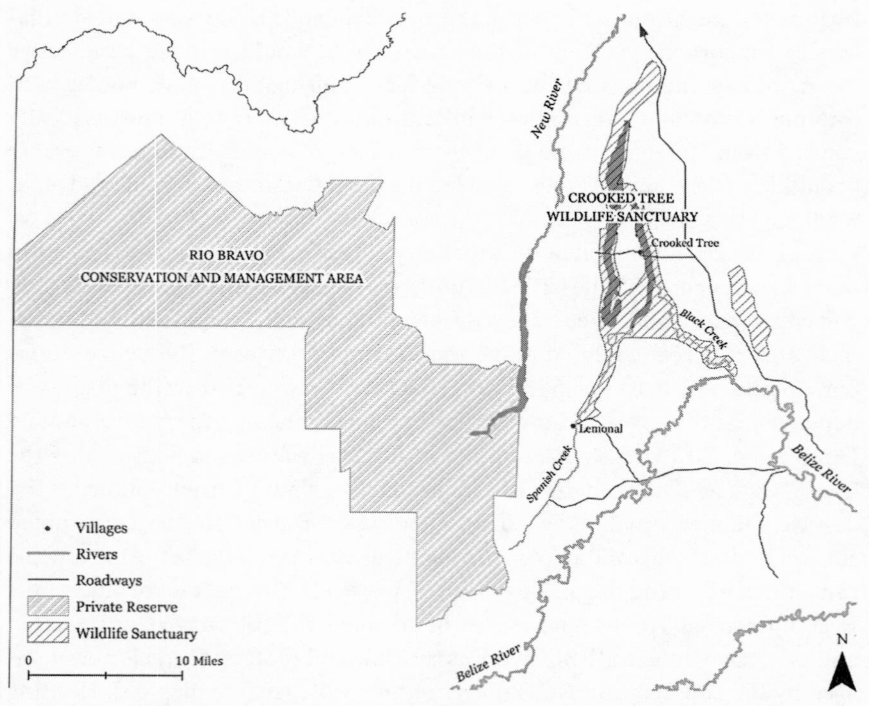

Map 6.1 Map of Protected Areas adjacent to Crooked Tree and Lemonal. Made for author by Caitlin Schneider and Simone Yoxall, in Anwar Sounny-Slitine's Geographic Information Systems Laboratory at Southwestern University.

Sanctuary, it was and is not heavily patrolled (Mitchell, Walker, and Walker 2017).

Mr. Elmore and the Duck Bag

Conservation efforts are predicated on an understanding of the inherent value of protecting the more than human, and the related idea that humans and human activity can be separated from birds, fish, mammals, plants, and landscapes. Dominant models of conservation also presume human activity to be of a different order than that of the more than human. Conservation is thus a moral act founded on Western ontological assumptions about the separation of humans from "nature" and the relative passivity of "nature" versus the agency of humans. As I described in Chapter 4, many rural Creole people see the more-than-human world very differently from this and do not necessarily fully share this particular moral vision. Rural Creole people live in ways very intertwined with the more than human, and the world beyond the human is agentive and powerful. Conservation regulations do not carry ethical and moral weight for many Creole people, but instead are a set of rules and regulations that are an annoyance to be evaded. This echoes rural Creole people's general sense of living outside the reach of the government. People living here have historically had a strong sense of being able to live as they wish, without having to worry about various kinds of laws (licensing, insuring vehicles, applying building codes to structures they construct, etc.).[8]

Mr. Elmore, a broad shouldered, brown-skinned, tall burly man in Crooked Tree (now deceased) articulated this stance toward laws restricting hunting very clearly to me. He was a renowned hunter and had worked in a variety of bush-related industries during the span of his long life, from mahogany to guiding trophy jaguar hunts. I was sitting with him in March of 1994, talking about his life and work in this part of Belize and about the Crooked Tree sanctuary. He said, "They have laws for everything, and you must respect the law. If you kill a duck and want to eat it, you have to respect the law, you put the duck in a bag and make sure they [the wardens, government] don't see you with it." For Mr. Elmore, the most important consideration is that one *must not get caught* breaking the law. Regulations and the possibility of getting caught exist, but the regulations have little moral relevance, and it is quite feasible to avoid being caught. This is what conservation constituted for Mr. Elmore and many others—an added effort, the possibility of being subject to fines and an annoyance. But, critically, conservation regulations did not stop people from hunting and fishing and did not fundamentally alter rural Creole people's entanglement with the more than human. There is also always a racialized dimension to these conservation assemblages in rural Belize. Laws, government, the state, and conservation all assemble more strongly with whiteness, on the one hand,

and living in these communities "off the main," or relatively out of the way, assembles more strongly with Creoleness, with blackness, and brownness, on the other.

Becoming Creole through Conservation

Not all rural Creole people reject, or sidestep, external logics as Mr. Elmore so clearly described in 1994. Indeed, some people living in these communities become entangled with conservation. For them, becoming Creole means becoming a conservationist, or thinking about limiting the use of the more than human in accordance with conservation biologists' and ecologists' reports and recommendations, at least for a while.[9] This is especially true for people who have worked as sanctuary wardens, or have worked elsewhere for Belize Audubon Society, Programme for Belize, other conservation organizations, or as nature guides for the tourism industry.

A story about a boy and a slingshot illustrates the kind of transitions that can occur. In 1990, when I first met then-seven-year-old Roderick, a beautiful brown-skinned boy with a broad smile, nothing delighted him more than making a good slingshot and killing as many birds as he could during the early dry season, when small songbirds, parrots, and parakeets congregate in cashew trees in the more bushy areas throughout the village. I always tried as hard as possible to be a good participant observer, to withhold judgment, to try to empathize and see the world through other people's eyes, but I found this slingshot hunting nearly unbearable from my white, New Jersey, professional, middle-class version of being a nature lover. One day, I could not help myself and I sharply reprimanded the boy for killing a few birds. He was "shamed" and furious with me (perhaps most of all for my having reprimanded him) and proceeded to shoot down even more birds in the following days.[10] His mother Miss Linda, a kind and gentle soul who had already raised three other boys, also discouraged him from killing so many birds. Her interests here were ambivalent. Having good aim and being able to hunt well have traditionally been important skills for Creole boys to learn, and any reasonable mother would be proud of a son who could hunt well. By the same token, as the matron of a family connected to the ecotourism business, she did not want tourists to see her son shooting down parrots, parakeets, and other birds. Her ambivalence expressed itself in the kind of casual way in which she chastised her son, and in the Creole child-rearing practice of allowing young boys freedom to move about the village. This same boy was also deeply fascinated by animals of all kinds and very interested in "bush" life in general.

When I returned three years later and saw this maturing ten-year-old, I was shocked. He had built a "killer" slingshot but now only aimed at fruit, branches, and fence posts. He told me proudly, "Ah no kill bud again. Da sin fi kill di bird dehn, yu bizniss fi lef di bud dey fi eat and live" (I don't kill birds anymore, it's a sin to kill birds, you should leave the birds alone to eat and live).[11] I asked him why he had changed his mind, and he told me that he had learned all about this in school. This was just one child at one moment, yet I have heard similar comments from other children over the years: "Da sin fi kill ting rite so" (It's sinful to kill things for no reason). An additional caveat: he only mentioned birds—it was still open season for game animals, jaguars, and fish. Several years later, by the time he had graduated from high school in Belize in the 2000s, he was a leading naturalist, using his skills in the natural world in Belize and the United States to make a comfortable living.

Roderick may have incorporated conserving nature into his rural Creole becoming, but mastering the use of slingshots to hunt birds and small animals is still generally part and parcel of growing up as a boy in rural Belize and an important step in gaining competence as a hunter and developing a sense of rural Creole masculinity.[12] Fifteen years after admonishing this young man for hunting birds for fun with slingshots, my own U.S.-born-and-raised Belizean American sons—eleven and eight at the time—visited their ancestral village of Lemonal. A friendly teenage cousin made each of them a beautiful slingshot and before I knew what was happening, my sons were becoming Creole—out hunting birds around the village with their cousins. This time I was quiet. I too was ambivalent—as I am about all hunting—but in the intervening years, I had also fully taken in the importance of this experience to becoming rural Creole for boys and men.

Tigahs and Cattle

The changes can also be more complicated and contradictory as they develop over time. In 1990, a young man, Darren, was working on memorizing English and scientific names for all the birds he had grown up knowing, so that he could be an excellent tour guide. Darren was clear-skinned and very fit, with a smile that could light up a room. By the mid-1990s he was a leading birding guide in Crooked Tree, and had so impressed some U.S.-based bird-watching tourists that they asked him to guide for them in the northern United States. With his success as a tour guide and naturalist, he had saved enough money to build himself a substantial cement house, but he was adamant that he would *not* "clean down" his yard like most villagers. He was determined to leave it bushy, to provide habitat for wildlife. He had also forsworn eating game meat, a very unusual stance in rural Belize. He was especially steadfast about tigahs (jaguars)—saying that if his father, who had killed his share of jaguars in the past, killed another jaguar, Darren would leave the village. Twenty years later, when he was in his forties, his stance had softened. Though still a highly successful nature tour guide, he had also begun working with his father tending cattle on their large ranch, and his children enjoyed a cleaned-down yard. Today, he is typical of many residents of rural Belize. He sees value in the restrictions on hunting and fishing and does not want the wildlife to "finish" (cease to exist, run out), so that tourism can remain a healthy industry in the area, but he wants to continue his rural Creole lifeways that sometimes do not accord with the goals of conservation science.

Other villagers have different thoughts. One important and highly respected leader in the community, Mr. Raymond, without whom the sanctuary may never have been welcomed, has always maintained a complex and ambivalent relationship to conservation and tourism efforts. Mr. Raymond agrees that there should be some restrictions on the haul of fish and on hunting, to ensure that populations can reproduce. He appreciates the value that the protected more than human bring to rural Belize in the form of tourism visits, and the money and pride those visits generate. But he also enjoys the variety in taste that game meats and fish provide, and he is a cattle rancher. Brown-skinned and wiry, Mr. Raymond is well known for his hard work and dedication to his village, his calm, unflappable demeanor, and the care he takes of his cattle—which sometimes fall prey to jaguars and pumas.[13] Belize has many of these highly endangered cats, and the area being described in this book likely has one of the densest populations of jaguars in the world.[14] Here, relatively large numbers of jaguars live in close proximity to large numbers of humans—in this case, humans with cattle ranches. Cattle husbandry is a historically deep Creole engagement with the more than human, and a marker of both Creole identity and elite status. Mr. Raymond, one of the largest cattle ranchers in this part of Belize, is well respected not only among the people who live here, but also by a variety of government of Belize agents who work with him. In 2012 and early 2013, eight of his calves were killed by a tigah.[15] Mr. Raymond set a trap, waited nearby, caught, and killed the cat. There is nothing terribly unusual in this—most cattle ranchers try to catch and kill problem cats. But in recent years the government of Belize, in conjunction with several different conservation organizations, has been more actively working to protect jaguars. Government officials have been more present and ready to confront ranchers and hunters who were killing problem cats. Furthermore, in the years just before this incident, wardens for the sanctuary had in general been more "facey" (in-your-face, imperious) in dealing with villagers, reprimanding some, fining others, and even confiscating a large quantity of logwood posts from one young man who was accused of illegally cutting.[16] Because conservation officials had been making more noise than usual, and because Mr. Raymond had lost so many calves, he brought the skin home with him and defiantly hung it up in front of his house, which sits on a main road through the village, for everyone to see. Someone reported this to the police and a representative from the Forestry Department visited Mr. Raymond at his home.[17] But because he is so widely respected, he was not charged or fined. Instead, the forestry representative scheduled a meeting for village cattle ranchers and government representatives to address the jaguar-cattle issue. Mr. Raymond described to me how a government representative framed the issue, and this framing spoke volumes. The official used the metaphor of the forest, the bush, as the jaguar's home, and then said that the trees were like the "jaguar's a/c." For Mr. Raymond, and for me as an analyst, this small comment encapsulated the outsider and elite orientation of government and NGO officials, who aim to implement conservation practices. Indeed, they are not very different from the colonial officers of one hundred years earlier. Mr. Raymond lamented that the government representatives seemed not to understand that the people in these communities are "poor and not rich," that they struggle to make ends meet. The tone-deafness of the "a/c" comment—very few people in the entire country of Belize have air conditioners in their homes—was remarkable and if anything served to reinforce Mr. Raymond's stance that the government had to meet villagers' needs. The meeting left the cattle-jaguar situation in Crooked Tree unresolved, but also meant in practice that ranchers would be left alone for a while.[18]

Halligata or People?

Conservation officials and their ways of communicating too often reflect elite perspectives and sets of concerns that do not resonate with rural Belizeans. But at an even deeper level, conservation policies can cast rural Creole people, Afro-Caribbean people, as an expendable category of human. The lagoons and waterways here are home to the Morelet's crocodile (halligata), one of the creatures that fully animates this landscape for rural Belizeans. Halligata were plentiful and common in the early twentieth century, but as demand for crocodile skins grew, so did commercial hunting.[19] Between the 1930s and 1960s, hunting halligata and selling their skins (either legally or illegally) was a lucrative endeavor for rural Creole people, and fit well with their kech-and-kill lifestyle. But by the 1970s, halligata had become scarce. The near extirpation of these animals prompted legislation to protect them, and in 1981 hunting halligata was outlawed. The market had also begun to dry. Crocodilian skin was not as popular as it had been, and alligator farms in the United States could readily satisfy demand. Belizeans stopped hunting them. When I first started traveling the lagoons and creeks of northwestern central Belize in the early 1990s, I rarely saw halligata—it was a big deal to see one. Each year I returned there seemed to be more. By 2016, the Morelet's crocodile was well recovered and flashlights at night on the waterways revealed blankets of shiny eyes (Platt and Thorbjarnarson 2000; Platt, Sigler, and Rainwater 2010; Platt et al. 2008). A robust trade in halligata meat emerged in the 2010s: Chinese-descended residents of Belize pay well to take home crocodile to cook. But while some hunters enjoy the new source of revenue that plentiful crocodiles provide, other rural Belizeans feel that the ecosystem is out of balance and that there are far too many crocodiles. For these people, the abundance of halligata serves as evidence of how misguided conservation efforts can be. A thoughtful and outspoken Crooked Tree woman, Mia, raised this issue during a conversation with me in 2013, one year after the halligata attack on the young fisherman described in chapter 4. She said that it is dangerous now to go fishing, to be on the water, in Crooked Tree because there are many halligata, and they are so big now. Mia was disgusted and distressed that nothing was being done to address this problem. For her and many other villagers, this is a failure on the part of conservation efforts, a valuing of crocodiles over and above rural Belizeans. Mia routinely hosts visiting college students from the United States, from schools that choose Crooked Tree at least partly because of the wildlife conservation initiatives here. Although Mia appreciates the benefits that wildlife conservation can bring, she does not welcome the ways in which these policies devalue certain human lives. Conservation and its handmaiden, nature tourism, bring places like Crooked Tree into the global mainstream, but being overrun by crocodiles generates a sense that rural Belizeans are only second-class citizens—that the imperatives of conservation outweigh the value of villagers.

Hicatee and Intensifying Entanglements

One other flashpoint for conservation in the lower Belize River Valley is hicatee, or the Central American river turtle. From the earliest moments, turtle has been a part of Belizean foodways. Hicatee is an important dry-season treat. Just as turkey graces U.S. dinner tables on Thanksgiving, hicatee graces Belizean tables on Easter Sunday. In neighboring countries, hicatee are extinct or nearly so; Belize still has a substantial number. Nonetheless, the hicatee population in Belize appears to be declining, and conservation policy has been established to redress this decline (Rainwater et al. 2012). The combination of declining stocks of the turtle and restrictions on hunting have in turn made hicatee especially "prized" in Belize's informal economy, which often does not follow legal prescriptions. Conservation efforts have arguably and ironically contributed to greater symbolic resonance for this rural Creole food item. Hicatee hunters can sell a turtle for a large sum of money, or might give one to a friend to repay a debt, or simply to connect to someone in the sharing networks that organize social life. For Belizeans living in the United States who return home at Easter, eating hicatee is more critical than ever to restoring themselves as rural Creole people. So, in this case, conservation efforts have intensified rural Creole socionatural entanglements in this place.

Outsiders, Race, and Conservation

If rural Belizeans were cautious and strategic when they first interacted with conservationists during the 1970s as the Crooked Tree Wildlife Sanctuary was established, they have only become more wary over the past thirty years (Johnson 2015). Rural Creole Belizeans often cast their relationship with conservation in racial terms. Indeed, conservation here frequently entails the presence of a phenotypically white person, and is facilitated with money and planning from international conservation organizations in the United States or Europe, which themselves reflect the whiteness of mainstream environmentalism (Garland 2008; Sundberg 2004; Taylor 1997; see also Nixon 2011). For example, in Crooked Tree, a white male Peace Corps volunteer was stationed in the village in the 1990s to develop a management plan for the sanctuary at the same time that my white self had recently completed a household census (Mackler and Salas 1994). During the months that this plan was developed, many villagers were skeptical of the work. Their comments were often framed in racial terms: "Wi no waahn no wite pipl fi kohn ya an tell wi weh fi do" (We don't want any white people to come here and tell us what to do). Although I was friends with most of the people expressing these sentiments, I became the focus of some *half*-joking hostility. I had supplied my census figure for the village to the author of the plan and my name was cited as the source of this figure. A number of villagers scoffed at and were incensed by these numbers, claiming that the numbers were far lower than the official count.[20] A well-known and well-liked man in the village, Dan, who at one and the same time was a prolific hunter, outstanding naturalist, tourism entrepreneur, and member of the board of Belize Audubon, weighed in on the management plan. From his life history and position, which itself captures so well the complexity and ambivalence of conservation and tourism in Belize, he decried the numbers I had provided and said that "Belize would become just like Africa," where whole villages were removed to create national parks for animals. He said that if people think Crooked Tree is small enough, "they will just move us off our land." "They" referred to government officials and conservationists (usually imagined as Belize Audubon), a "they" that also coded whiteness.

I unwittingly participated again in the whitening of conservation in the months before the plan was drafted, and at the beginning of my long-term research period in Crooked Tree. There was supposed to be a meeting between Audubon and village council officers to discuss the regulation of fishing. It turned out on this day that the officers had other urgent business and could not attend the meeting. However, Audubon's Mestizo manager of Protected Areas and a white American woman who was working as a consultant with the Department of Conservation had driven several hours for this meeting, and the two of them decided that they should at least hold some kind of informal meeting. The two Creole wardens from the sanctuary and a Creole man who was a leading figure in the tourism industry were present from the village. I was there with my white self, technically merely as an onlooker (as I had requested to be). Initially, the Protected Areas manager acted as facilitator, but the meeting was quickly dominated by the white consultant, and I soon found myself feeling very uncomfortable. Whenever a question about the village needed answering, or some aspect of village life needed clarification, the consultant, and to a lesser extent, the Protected Areas manager, looked to me for answers. I knew that both the wardens and the tour operator knew much more about the village, about their village and themselves, than I could ever hope to. But the consultant did not turn to them when she had a question. Her exclusion of the villagers was probably not intentional. She was likely more comfortable talking with me, our shared whiteness binding us two strangers together, and with her friend, the Protected Areas manager. And this was very likely only one among many similar situations the villagers had witnessed—people from "foreign" and white people talking to each other about rural Creole Belizeans, in front of rural Creole Belizeans, unintentionally and implicitly discounting and excluding them.

The sensibility that conservation is a set of policies and regulations originating far from rural Belize also marked a meeting in Lemonal in 2013. The Central Belize Corridor, a project of the University of Belize's Environmental Research Institute (ERI), aims to address the problem of habitat patchiness for large mammals, like the jaguar, puma, peccary, and tapir, that need expansive stretches of habitat to thrive.[21] The corridor runs through northwestern and central Belize, tying a series of large protected areas together. In 2013, a small group of ERI representatives held meetings in the different villages located within the corridor to explain the project, and to begin a process by which rural people could identify how they want to engage with the project: how they want to manage the lands and resources where they live. Each village was to draw up its own management plan. The meeting in Lemonal on a Saturday afternoon in June attracted representatives from many but not all households

Belize is a relatively weak state and its conservation sector is underfunded. This context in conjunction with an international emphasis on participatory processes in conservation practice meant that the corridor representatives were hoping to initiate a participatory process with this meeting.[22] This was their first meeting with the people of Lemonal, and their first effort to bring these villagers in to participate in helping create the corridor. However, from the perspective of Lemonal villagers what was being offered here was no different from previous efforts at conservation. One older man recalled his time living in the United States and declared that the United States and other big countries had destroyed their environments and now they want to come to the small countries and have the environment to themselves—nothing comes of this for countries like Belize, he said.

Even though the representatives were young Belizeans (Creole and Mestizo) and the program emerged out of the University of Belize, it was still fundamentally seen by villagers as something from the outside. A well-respected village leader, whose children have attended the University of Belize, articulated a sentiment that underlies many rural Creole people's feelings about conservation. She said that the people here in Lemonal and other rural villages do a good job of keeping the place "environmentally rich."[23] She noted that a bigger problem is that many people do not have official title to the lands to which they have long ancestral connections. Some wealthy person could come, buy the land out (from the government), and then destroy the area. Villagers nodded in agreement. She and others at the meeting made clear that the way they live with their land and environment is not only critically important to them but also sustainable. These communities have been in these places for hundreds of years and remain among the most ecologically rich in the country.[24]

This meeting, which was similar to many other meetings I have attended between conservation officials and rural Creole people, demonstrates yet another dimension of how conservation assembles with rural Creole socionature. Conservation always carries with it a brush of "outsider," "expert," "white," and "colonial." In addition, conservationist discourse and practice too often cast rural Creole people's ways of being—hunting and fishing entangled with the more than human around them, as something wrong, as a problem. From the perspective of conservationists, rural Belizeans and their socionatural becomings stand in contradistinction to Belize's newfound fame as a star in the international conservation arena. But for rural Creole people, their long history of living in these environmentally healthy lands offers proof that Creole socionatural entanglements and Creole attachment to place are sustainable and should be celebrated by conservationists.

While individual rural Creole people might incorporate conservation ethics and practices into their lifeways or reject them, conservation itself is an assemblage that excludes rural Belizeans. If becoming rural Creole means becoming with the more than human in these tropical lowlands, many forms of those becomings are not enfolded into conservation, even as conservationists claim to want to include rural Belizeans in their projects. Conservationists want to include a hollowed out, diluted version of being rural Creole.

Conservation is like other external logics that rural Creole Belizeans have engaged over the past two centuries. Belizeans have multiple and changing modes of interacting with conservation, using it to their advantage, absorbing and incorporating it into their socio-ecological modes of being, and/or standing in opposition to it. All these engagements are also simultaneously a part of processes of racialization—conservation's associations with whiteness, outsider status, and the elite are always a part of how rural people engage these initiatives.

The Natures of Tourism

The global project of tourism circulates into rural Belize tightly tied to conservation, but with slightly different effects, generating different types of encounters and relationships that in each moment bring into relief difference: racial differences, different positions in global cultural economies, different ways of being in the world, and different socionatures. These encounters have the potential to either strengthen or weaken dominant associations. Contemporary nature tourism in Belize builds on colonial sport and trophy hunting and fishing, which gained popularity during the first half of the twentieth century throughout the British Empire (Hannam 2007; MacKenzie 1997). British Honduras was an exotic terrain that attracted Americans and Britons as an exciting and unknown place to explore in the early twentieth century.[25] Its healthy populations of large, dynamic sport fish, like tarpon, and the presence of romantic and elusive jaguars, as well as mysterious Maya ruins all served as key attractions. The lower Belize River Valley, Crooked Tree, and New River Lagoon figured centrally in these early connections.

Early planning for tourism and conservation were brought together in the National Parks Study Group in the 1950s. This study group was made up of white colonial officials and expatriates from England and the United States.[26] They worked to identify beautiful areas in the country that would attract tourists (and not just fishermen and hunters). Crooked Tree with its lagoons and unusual and plentiful birds was on the list of the initial fifteen places recommended for protection. The anticipated tourists were the growing numbers of predominately white professional middle class in the United States and Britain. The very establishment of a protected area in Crooked Tree was made appealing to villagers by the possibility of visiting tourists and the ensuing opportunities for income and employment villagers could expect.

With Crooked Tree Wildlife Sanctuary and a small number of other early nature-tourist destinations (including the coral reefs of the Cayes), as well as a large percentage of relatively unbroken tropical forests, Belize became a hot spot for nature tourism starting in the 1980s. Costa Rica may claim the mantle of being the primary ecotourist destination in Central America, if not the world, but Belize has never been far behind, calling itself the "jewel of the Caribbean" or just "the jewel." Ecotourism thus allowed Belize to claim a significant role in the global cultural economy (Medina 2012; Wilk 2006). At the same time, reflecting the ambivalence regarding the bush that I described earlier, Belize's pristine environment is a result of, and a sharp reminder of, Belize's longtime status as a colonial backwater. By 2007, reflecting international trends in the global growth of tourism (Mowforth and Munt 2015),, tourism contributed to 25 percent of the jobs in Belize and made up 18 percent of the country's gross domestic product (GDP) (Barrow 2008). In 2012, estimates suggested that tourism directly and indirectly contributed to 32 percent of GDP, and that it was the largest economic sector for this country of less than 350,000 people. Furthermore, tourism continued to grow in Belize in the second decade of the twenty-first century, despite declining tourism in similar Caribbean sites (Nuenninghoff et al. 2014). By 2016, tourism's direct and indirect contribution to GDP had risen to 38.1 percent (World Travel and Tourism Council 2017).

Rural Creole people had experience with the precursor to the ecotourism industry—sport fishing and hunting services for wealthy white men, usually from the United States. Some worked in several-month jobs as guides, cooks, or housekeepers at Keller's Fishing Lodge—a lodge near Ladyville on the Belize Old River established in the 1960s by a white man from the United States and still in business under a different name today—specializing in tarpon and other sport fishing. Others worked as guides for short stints in the 1960s and 1970s for a trophy jaguar hunting camp that released captured jaguars for wealthy U.S.-based hunters and thus guaranteed each paying customer a trophy jaguar. This hunting camp was run by a U.S.-based white man who later ran similar hunts at a camp in the United States.[27] A few individuals worked at the sport-fishing-oriented Spanish Creek Lodge located a few miles downstream from Lemonal on Spanish Creek, which operated in the 1960s and 1970s.[28] These Belizeans experienced a range of relationships and treatment in these positions. Sometimes, mostly at the fishing lodges, the work was well paid and the Creole people who held jobs felt they were well-treated. But the Creole men who worked at "tigah camp" were poorly paid and treated disrespectfully and with overt racism. In many ways, rural Creole peoples' work in these jobs was not hugely different from their prior experiences working in the timber industry (see Büscher and Davidov 2013): some people had good jobs, some had not-so-good good jobs, and they might experience either respect or racism at the hands of their typically white employers, depending on the situation. These experiences, shared as stories within rural Creole communities served as the backdrop for the development of tourism in these places.

Tourism had been steadily growing in Belize when I first visited to do research in 1990. In the early 1990s, ecotourism in Crooked Tree was at its early stages. Today the village boasts several resorts that provide accommodations, meals, and guiding services, as well as bed-and-breakfast services and a few restaurants. Tourism businesses in the village are all owned and operated by people from Crooked Tree, and many households in the community have some connection to revenue from tourism. The road to Lemonal village goes through the well-visited tourist community of Bermudian Landing, home to the Community Baboon Sanctuary, which boasts a friendly and interactive group of wild howler monkeys and a number of tourist accommodations.[29] In 2016, the road into Lemonal was paved, at least in part because of tourism development. The road was extended out into the bush to the north of the village to a new embarkment point for cruise ship tourists to take boat trips across New River Lagoon to see the Maya ruin, Lamanai. Several villagers have found employment with this new business, which is owned and run by someone from another community in the lower Belize River Valley.

Nature tourism is today part of the warp and weft of everyday life in Belize, and touches everyone who lives in the nation. Becoming rural Creole now always includes ecotourism—having the more than human among whom one lives on display and exchanging for money knowledge of the natural environment and skill in navigating these tropical lowlands. In many ways, working in nature tourism is a very logical development for those who have long made livelihoods from this place, both as people sustaining themselves through relational socio-ecologies and as wage laborers in forest industries. The patchiness of tourism-related income fits well with the occupational multiplicity (Browne 2004; Comitas 1973) of Afro-Caribbean rural Belize, and facilitating ecotourism constitutes another way of becoming with the creeks and lagoons, tubroos and cedar trees, jabiru storks, iguana, and other more than human here. Tourism has affected these entanglements, emphasizing some skills and abilities more than they might have been emphasized earlier: spotting birds and other animals, navigating and maneuvering outboard-motor–powered skiffs through the sinewy waterways, for example, are critical to the success of nature tourism here.

Globally, nature tourism is also part of socionatural assemblages in which racial difference emerges prominently. Margaret Werry (2011), Robert Fletcher (2014), and Bruce Braun (2003) have shown how different forms of ecotourism are intimately tied to the racial formation of whiteness (see also Finney [2014b] and Outka [2008] on blackness and wilderness use in the United States). The prototypical tourist is of European ancestry, phenotypically white, and relatively elite whereas many ecotourism sites are located in the tropics, home to darker-skinned peoples, or in out-of-the-way places where nonwhite indigenous peoples live. In tourism encounters, these differently embodied individuals come together in particular places, with the more than human and with a host of racial codes, or symbols and signs that tie to specific racial projects. In any given encounter lies the potential for racial formations to become more or less solidified.

The whiteness of the ecotourist often includes an unspoken sense of entitlement, as was illustrated by the Texas duck hunters in the 1970s in Crooked Tree. A thought experiment is illustrative: white, upper-middle-class tourists from the United States walk through rural Belize differently from the way one might imagine rural Belizeans walking through a wealthy white New York suburb, for instance. Indeed, such a suburb would *not* have tourists walking through the neighborhood the way tourists walk through rural Creole villages. Residents of rural Belize have been savvy about how race works in tourism from the outset, as indicated by some of the stories I share above. Notably, early advertising developed by Crooked Tree villagers described the area as populated by "descendants of English and Scottish woodcutters," with no mention of the village's history of slavery or African ancestry. In these brochures, they claim and celebrate their centrality to Belize's national origin story in a way that is most palatable to white tourists.

While entitlement and other racially coded sentiments and discourses assemble with the whiteness of the ecotourist, racialization is not fixed and set in stone, but always has the potential for assembling differently. Race assembles through these encounters, as place, social class, water, birds, mammals, reptiles, insects, knowledge about these more-than-human entities, desire for authentic nature, and more come together with differently embodied persons. This is not to say that there is no dominant global racial order, just that possibilities for something *otherwise* are always also present.

On Being Nature Guides

When rural Creole men parlay their deep knowledge of the waters, winds, birds, fish, plants, mammals, reptiles, and other more-than-human elements of this place into work as naturalists and tour guides, they continually generate Belizean socionatural becomings. These becomings assemble blackness and brownness with knowledge, competence and skill.

Contrastingly, tourists traveling from the Global North, bring with them racializing assemblages that constitute whiteness as simultaneously normative and invisible on the one hand, and discursively associated with an implicit imagined superiority in a range of arenas, none more significant than rationality, knowingness, and intelligence, on the other (Bonnett 1999; Goldberg 2001). Indeed, these associations, along with the necessarily related pathologizing of blackness and brownness, are the pillars of what Hughey has called "hegemonic whiteness" (Hughey 2010). When a tourist from the whitened Global North travels to blackened African diaspora communities in the tropics, these associations constitute the white racial frame (Feagin 2009) through which the tourist experiences this encounter (Braun 2003; Munt 1994). Thus contrasting racializing assemblages confront one another when a black or brown Afro-Caribbean ecotour guide sells his nature- and bird-guiding knowledge to an ecotourist. I share below one of many ethnographic examples I have that illustrate how blackness, whiteness, and, in this case, masculinity, convene in this kind of encounter.

In the late 1990s, a white male German tourist, Edward, was looking for accommodation and guide services in Crooked Tree. An avid birder and naturalist, he wanted to go on a boat trip through the lagoon system to see waterfowl and other wildlife. He booked a room in the lodge owned by Thomas, a rural Creole man I know well. Edward asked Thomas what the cost for such a trip would be, and expressed shock at the price he was quoted. Prices in Belize tend to be higher than in nearby Mexico and Guatemala, so some surprise is not unexpected. However, the way in which Edward haggled with the lodge owner contained tones of disrespect, of incredulity that the Creole guide to be hired, a young man, John, would offer Edward anything more than knowledge of how to run a boat down the lagoon. Thomas, having experienced this kind of questioning of the value of rural Creole guiding services many times before, was clearly annoyed and made a final offer, which Edward, while still complaining, refused. Edward simply could not imagine that what John would offer would be worth the money, and was convinced he was going to be swindled. Hegemonic white frames that preclude the association of blackness with deep knowledge and expertise contributed to the way in which Edward navigated this encounter. If Edward had been visiting an Audubon center in the United States, with its schedule of prices to hire a (white) birding guide, he would have been less likely to haggle so adamantly. Lodge owner Thomas walked away. Edward returned the next day, broke down, and agreed to pay the guiding fee. The following morning, at daybreak, Edward climbed into a motorboat with John, the young Creole man who would be his tour guide, naturalist, and boat captain. Edward started his trip defeated and resigned, assuming he had overpaid for little value, and his haggling had soured Thomas's mood as well. But when Edward returned from his daylong birding trip in the evening, he was transformed. He excitedly exclaimed how utterly amazed he was by John's knowledge of the place, the birds, the animals, and plants, and the ecological relationships between all the inhabitants of the lagoon and river complex. Edward could not praise the guide enough, and exclaimed that he had gotten so much for so little. Indeed, he was so pleased, he happily paid the fee that had seemed exorbitant to him the day before, to go out again on the following day with the same rural Creole guide.

Edward's dogged haggling in the early 1990s was not an uncommon occurrence, and probably also reflects Belize's relative newcomer status to the international tourism scene at that point in time. In the intervening years, guidebooks for touring Belize celebrate the excellence of the nature guides found throughout the country. Nonetheless, tourists are still surprised at the intellect and expertise they experience from their Afro-Caribbean guides. For example, in 2013, bird-watchers from the United States hiring guiding services at a different lodge in Crooked Tree exhibited a similar set of low expectations, if not so blatantly expressed, and also exuberant surprise at the guide's expertise. My half-black Belizean and half-white American family were staying at the lodge ourselves, enjoying a tourist experience in Crooked Tree (see figure 6.2). We went on a guided birding and wildlife-viewing trip through the lagoon and down Spanish Creek. This was an interesting experience for all of us. My husband

Figure 6.2 Author and sons Elrick and Adrian Bonner at tourist lodge in Crooked Tree, 2013. Photo by Elrick Bonner.

Elrick is well known throughout the Creole villages of Northern Belize for being an excellent hunter and for being very skilled at spotting birds and animals. Elrick and I both knew the guide, a young Creole man in his twenties, whom I remembered best as a cherubic child from a decade earlier. The guide did an excellent job, even if Elrick spotted one or two of the crocodiles before he did. My family and I thoroughly enjoyed the trip, and learned a lot. That evening we were having a meal in the lodge's restaurant. A white couple, bird-watching tourists from the United States, were seated at the table next to us, and we started chatting about how we had been spending our time at the resort. When we spoke of the birding trip we had taken, they shared that they had gone on the same kind of trip (with the same guide) and had also gone on trail hikes with another rural Creole man who showed them birds and wildlife. They exclaimed how "absolutely amazing" their guides had been and how very surprised they had been by this.

This subconscious devaluing of African-descended people's expertise and capability is of course foundational to the origin and perpetuation of antiblack racism and to the elevation of whiteness that marks global white supremacy. Belizean Creole people's experiences of having their selves and labor undervalued traces back to the earliest days of the colony. But being underpaid for work that requires deep knowledge of and skill in negotiating tropical lowlands is doubly dehumanizing. As I have argued throughout this book, this knowledge and these competencies are integral to how one becomes Belizean Creole. They are central to people's understanding of themselves. Thus, the process of racialization that denies them their intelligence tied to this place also strips them of the culture that makes them who they are. Rural Creole reaction to this belittling is bemused indignation: bemused because of the tourists' ignorance and white people's general softness and incompetence, indignant because of the implication that they and the place of Crooked Tree are somehow worth less, occupying an inferior slot in the global scheme of things.

Yet these moments also serve to destabilize hegemonic whiteness and its concomitant pathologizing of blackness (Hughey 2010). The tourists' surprise at the extraordinarily high quality of their black guide's knowledge and expertise, including mastery of Latin species names and the complex ecological relations of these subtropical ecosystems, challenges the subconscious racial frames that organize tourists' understandings of the world. The growth of tourism in rural Belize has also allowed villagers to see their detailed knowledge of the natural world increasingly valued in the global political economy. In the nature tourism and bird watching sector of the global tourism industry, rural Belizean expertise in and knowledge of the more than human has become sought after. Not long after Crooked Tree became a celebrated location for birding, several young men from Crooked Tree who were particularly skilled and knowledgeable guides, including John from the anecdote shared above, began being invited to apply their skills elsewhere. Birding and nature-oriented summer camps in the northern Midwest region of the United States have hired men from this community, one guide was invited and paid to assist on a safari in East Africa, and another guide now routinely leads birding trips throughout Belize, Central America, and other parts of the Caribbean.[30]

Food and Ecotourism

As I described in chapter 3, certain foods are seen as "bushy," uncivilized, far from urban, and, I argue, are also racialized as distant from whiteness. Yet it is typically those bushy foods like wari (a type of peccary), gibnut, and hicatee that are some of rural Belizeans' favorite foods. These are "fi wi" (our) foods, foods that embody being rural Creole. When Belizeans run tourist lodges, restaurants for tourists, or hold festivals that might attract tourists, the question of what food to offer engages discourses of bushy-ness as "backwaad," on the one hand, and rural Belizean cultural pride, or pride in Creole socionatural entanglements, on the other. Although Belizeans are proud of the variety of game animals they hunt and the skill they have in cooking them, these wild meats are rarely sold in restaurants for tourists. Tourists may indeed imagine themselves wanting to experience the local ecology closely, but the typical ecotourist is the product of a culture that imagines humans and nature as separate entities. The tourist is happy enjoying the beauty of the landscape and its animals, but does not want to have to eat one of those animals for dinner. The ecotourist wants to go to the altar of nature, while the villager is more content with the nature at the kitchen table. The foodie movement and the increasing interest in exotic and unusual foods (see Wilk 2006) may be changing this, and Belizeans may be more encouraged to share their unique rural Creole foods with visiting tourists.

Reprise: The Flopping Fish

Tourism thus has the effect of emphasizing and valuing only some aspects of rural Creole people's entanglements with the more than human. Their knowledge of the bush, the animals and plants that live there, and the life habits of different creatures, are all celebrated by tourism. Other aspects of rural Creole socionatural entanglement are not only *not* embraced by tourism, but openly rejected by tourists. Encounters that feature these rejections reveal the ambivalence and complication that ecotourism entails for rural Belizeans.

The anecdote I share at the beginning of the chapter, describing a Belizean teenage boy, his flopping fish, and an older white bird-watching couple all sharing a pier over the lagoon one evening in the 1990s well illustrates the kinds of dynamics at play. Nature tourists, who are often white middle-class suburbanites from the United States, are typically most comfortable with only sanitized versions of the more than human, where the blood, guts, and suffering that constitute everyday life are not visible. Rural Creole people are aware of this gap in the way that white tourists see and what they are capable of handling. White people are seen as soft and unable to deal with the challenges posed by "life da bush." The teenager's chuckle and his persisting in Creole fishing practices on the pier are assertions of rural Creole emplacement and lifeways and a public expression of determination not to be fundamentally changed by tourism. His chuckle was also about his pride in himself, valuing who he was, positively assembling brown skin, water, fish, competence, skill, and toughness-in-living in this bird-watcher's paradise.

Conclusion

Conservation and tourism join the assemblage of rural Creole people and their socionature. While they have clearly affected everyday life in this place, rural Creole people persist in their own ways of making livings, ways that are not fully determined by the global political economy to which they connect. They continue to engage a powerful and agentive natural world that is not recognizable to conservationists and tourists.

These rural Belizeans have used conservation projects to further their own interests, including controlling what types of white outsiders are allowed in this place, and otherwise incorporating or rejecting conservation's logic, but always engaging with conservation in tandem with pre-existing modes of socionatural entanglement. There is an ongoing sense of pride in Creole competence and a commitment to maintaining the socionatural assemblages that make a person Creole—fishing, hunting, recognizing the more than human as agentive, exchanging the more than human in circuits of reciprocity. Present-day conservation efforts are part of a long history of what Creole people see as an unwelcome external intervention into local ways of being, typically not a very successful external intervention. Nature tourism's close ties to conservation make it difficult to analyze one without considering the other. Belizeans welcome the opportunity to show off their communities as well as the tourism-related jobs that very much fit within a rural Creole way of making a living. While some analyses of conservation and tourism development in different parts of the Global South emphasize how conservation and tourism generate major shifts in livelihoods, economies and subjectivities, I argue that this is not the primary story for Crooked Tree, Lemonal, Bermudian Landing, Double Head Cabbage, Flowers Bank, Rancho Dolores, Biscayne, or any of the other communities of the lower Belize River Valley. Rural Creole people continue to create socio-ecologies that conform mostly to their own sense of timing, history, and being. They create economies and ecologies *otherwise* (Gibson-Graham 2008, 2011). Being rural Creole now also includes working as a tour guide or wildlife sanctuary warden, and for the occasional individual, no longer eating game meat. But being fully entangled with the more than human continues to be how one becomes rural Creole, and indeed, conservation efforts have sometimes deepened the connection between becoming Creole and the more than human, as in the case of the Central American river turtle—or hicatee.

Tourism and conservation repeatedly entangle rural Creole people into processes of racialization that are complex and contradictory. White tourists' often low expectations of guiding services are frequently confounded, while Creole people are reminded of their competence and knowledge in tropical lowland ecologies. Conservation practitioners and projects continually "low-rate" rural Creole ecological knowledge. These reckonings assemble with the phenotypically white, brown, and black skin colors that are part of conservation in ways that sometimes stabilize racializing assemblages (conservation tends in this direction), or loosens and re-assembles race differently, as tourism is more likely to do.

CHAPTER 7

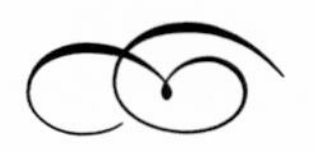

Transnational Becomings

FROM DEER SAUSAGE TO TILAPIA

One morning in early fall 2010, as I sat down to write, I worried about my to-do list. A box containing corned deer meat that had been salted and hung to dry in the hot Texas sun was sitting in the laundry room, waiting to be fully packed and shipped to one of my four sisters-in-law in Chicago, the sister-in-law where my then seventy-four-year-old father-in-law was staying. These women had left Belize in the late 1990s and early 2000s, part of a steady stream of migrants from Belize to the United States. They joined the large community of Belizeans living in Chicago, where pre-existing networks offered well-paid domestic work for women. Like thousands of other Belizeans in the United States, they were waiting for "their papers to come up," or the opportunity to adjust their immigrant status to permanent resident through a relative (in this case, my husband Elrick, who himself had become a citizen through me). While waiting, they could not leave the United States. Many Belizean immigrants stay in the United States longer than their visas permit, and are undocumented. A visit home to Belize would mean abandoning their life and family in the United States, where they had children, well-paying jobs, and large social networks. So they wait and in the meantime look forward to all the different ways they can find to replenish their Belizeanness, their ties to particular places and communities in Belize.[1] For my sisters-in law and their families, corned deer constituted one of these ways. Elrick had prepared the meat three weeks prior and was waiting for me to help with the final packing and delivery to the post office. The deer was hunted in Central Texas by a white friend of our white rural-north-Texas-born-and-raised neighbor, who is a frequent fishing and hunting companion of my husband. The freshly killed deer was brought to our house, where Elrick butchered and cleaned the animal in our backyard, Belizean style—the carcass hung from a tree limb, skinned, bled, and then cut into pieces with plentiful bones, the way most Belizeans prefer their meat. The final step of salting and curing in the sun fully rendered this Texas deer into Belizean meat. I wondered if this meat would suffer the same fate as the last batch we mailed northward to family (sisters, brother, father, cousins, and other kin); meat intended for father, but greedily and happily taken by one sister, causing amusement and vexation all around, and fully part and parcel of the networks of giving that generate a sense of being Belizean, being Creole, being rural, being a member of this family that is so deeply rooted in rural Belize. The corned deer awaiting shipment assembled together white rural U.S. southern hunting traditions, a white-tailed deer, the hot Texas sun, Belizean Creole people, and their movement in Belizean networks, and these assemblages replenished Belizean Creole identity. This assemblage aptly illustrates the twenty-first-century mobilities that characterize rural Belizean socionature, highlighting the different possibilities that might emerge out of these moments. The anecdote neatly came full circle six years later. During an online video chat, a boyhood friend and cousin of Elrick's still living in rural Belize shows him a big bag of deer meat and promises to corn a piece for Elrick when he visits Belize in two weeks. The offer was a gesture toward rebuilding what had been a strained friendship in the years before.

These circulations of human and deer contribute to always-becoming socionatures, both in the United States and Belize, and in both places generate afresh what it is to be a rural Belizean from these villages. I have already described how rural Belize Creole socionatural becomings have been entangled with extractive timber economies in the nineteenth century, and in the twentieth and twenty-first centuries with conservation and tourism. Here I focus on another global process that rural Belize engages: migration, or the movements of both human and more than human across borders.

Thinking transnationally opens a new way of understanding the relationship between people and their more-than-human entanglements. Most studies of how human relationship with the natural world creates culture and belonging, recently most commonly encountered in the multispecies turn, focus on one particular place (e.g., Blaser 2010; De La Cadena 2015; Pilgrim and Pretty 2010). In this chapter, I consider how the more than human in Belize *and* in the United States both are enfolded into becoming rural Creole, and how animals and plants that move transnationally enter new and unexpected socionatural assemblages. Socionatural becomings span across places and nations (Braun 2006).

Becomings with the more than human that generate rural Creole belonging occur not only in rural Belize, and they do not always only include game meat with a rural Belizean provenance. Belizeans living in Central Texas, Chicago, New York, and Belizeans who move between Belize and North America, also engage practically with the more than human, sometimes with plants and animals that themselves are from Belize, sometimes with plants and animals that are from North America but have been transformed into Belizean socionature, and sometimes with animals that originally come from even farther way, such as tilapia that come from Africa. I argue that these entities assemble to become Belizean socionature, and that these becomings occur throughout rural Belizean translocal networks.

The Caribbean has been described as "mobile, fluid and dynamically articulating" (Thomas and Slocum 2007, 3) across many types of boundaries; its diaspora as central to its definition as the archipelago that conventionally defines the region. Indeed, contemporary societies in the Caribbean, such as Belize, have only come into being through processes of global movement and interconnection, and these interconnections have always, since they first happened in the sixteenth century for the Americas, entailed the mixing of peoples and material goods (Wilk 2006).[2] Being from the Caribbean means, as Mimi Sheller notes, "Being uprooted from one place and re-grounded in another such that one's place of arrival becomes a kind of reinvented home. It implies the displacement (yet not total loss) of a previous home/culture and the claiming of a new belonging" (Sheller 2003b, 276). Violent uprooting, and re-rooting in contexts of racialized structural inequality, brought the Caribbean into being. Displacement, up-rooting, and re-rooting continually recreate Afro-Caribbean communities in diaspora in the United States and beyond.

The transnational assemblages that constitute identity and place at any moment and in any location emerge out of connections to places of origin, new places, and places of the imagination (Brenick and Silbersein 2015; King and Christou 2011; Malkki 1992; Salazar 2011; Santiago-Irizarry 2008; Smith and Guarnizo 1998). For Afro-Caribbean people, who have been "homing freedom" (Crichlow and Northover 2009b), these place-making processes, this reassembling in which they take part, although situated within lines of racialized inequality (and violence), often provide space and room for freedom, for creative potential, for living *otherwise* (Gibson-Graham 2011). Often overlooked, these quotidian modes of being, in backyards and kitchens, are powerful sites for the formation of cultural identities in entanglements with the more than human (Morgan, Rocha, and Poynting 2005, 105). Indeed, I argue that in yards and kitchens in Chicago, Waukegan, Austin, and Houston, migrants from Belize replenish their Belizeanness and raise Belizean (American) children by entangling people with fruits, meats, and fish, by practically engaging the socionatural world. These cultural sensibilities become further strengthened by return visits, or longer-term relocations in Belize. And all these processes become intensified through social media. The socionatural assemblages that make us human, that make people rural Belizean, are thus multi-sited. Tracing their linkages to emerging identity illuminates how "cultures of migration" are generated (Cohen and Sirkeci 2011).

Sharing beyond Borders

Being rural Creole Belizean means being entangled with the more than human and being competent in the natural world—knowing how to hunt, fish, find particular fruit trees and other goods from the bush, knowing how to prepare all these hunted and gathered foods, knowing the landscapes surrounding the community in which one was raised, and relishing eating particular noncultivated and nonimported foods. A sense of Belizeanness and community is also continually generated through networks of sharing, including the sharing of foods, and particularly game, fish, or food obtained through work in the natural world, and not foods purchased from stores and shops. These skills and practices confer a sense of pride in being "haad" and able to smoothly handle the difficulties that "life da bush" presents.

Hunting, fishing, and preparing and sharing these foods from and in this place are thus central elements of the socionatural assemblages that generate being rural Belizean Creole. Yet, over half of all Belizean Creoles now live in the United States, the majority in Chicago, New York, and Los Angeles (Woods, Perry, and Steagall 1997). As the story about corned deer illustrates, rural Creole Belizeans living in the United States reproduce knowledge, skills, and socialities in new contexts when they can—hunting and fishing when they can find places to do so, transforming the meats and fish they procure from U.S. soils into Belizean foods, and then putting these into circulation among social networks to (re)produce Belizean social relationships.

The corned deer story that I opened with provides one such example, but my extended family has been a part of many similar experiences over the years. Game meat that is hunted or fish that are caught are then prepared in a Belizean Creole way and shared with Belizeans living in the United States. Sometimes the food unites Belizeans born in different places throughout Belize and from different ethnic groups (Garifuna, Mestizo), such as in Central Texas where there are fewer Belizeans generally. In Chicago, where there are large numbers of migrants from each village in Belize, the networks are more likely to be primarily from one or two villages, like Crooked Tree and Lemonal. These wild foods thus simultaneously broadly link Belizeans together but they also produce an intensified sense of rural Creole Belizeanness and connection to one's home village.

For example, when Elrick travels to Chicago the airline trip to the windy city will include a cooler of frozen deer, wild pig, javelina (or peccary), and/or fish, traveling as luggage. The meat and fish are carefully shared—siblings, close relatives, dear friends, and others who might have been in telephone or online conversations and said they wanted to "taste" the meat will be sure to receive some. The recipients relish this wild flesh, which tastes a little more like home than the grocery store fare in Chicago, and is cut in ways better suited to

Belizean cooking than the standard cuts of meat and fish found in U.S. grocery stores. Having these foods allows migrants in the United States to satisfy an embodied longing for home—for the smells, tastes, and sounds that are life in rural Belize (Dundon 2005).

Because we are one Belizean family among a handful from Lemonal and Crooked Tree networks who live in rural parts of the United States, we more often ship than receive meat and fish. But we nonetheless still receive parcels of various foods shipped to or shared with us—and the ones we value most are the ones that Belizeans have wrested from the earth and produced themselves. Late in the fall of 2004, for example, our friend Ed brought us "cuttobrute" (made from coconut, butter, and sugar), and "wangla" fudge (made from wangla, or sesame seed, and sweetened condensed milk) that he had received as a gift from an aunt. Ed explained to us that his aunt had used coconuts from her yard in a rural village and wangla from her husband's plantashe to make the treats that we received. That the foods were from a village yard and prepared by someone connected to us made them that much more valuable and strengthened the tie between Ed and us.

Sometimes freshly caught game and fish are brought from Belize into the United States by traveling Belizeans. Hicatee and the river fish crana and bay snook are popular, as they are powerful symbols of rural Belizean living. Belizeans living in the United States go to great lengths to taste these foods. For example, in April 2015, a promised treat of fresh hicatee in Houston prompted a group of about ten Belizeans living in Central Texas to plan a weekend drive to the "oil city" (about three hours away). At the last minute the hicatee dinner did not make it through customs at the Houston airport. The travel plans to Houston were abandoned and the Belizeans disappointed. When turtle, fish, or other game meat are successfully brought into the country, they are carefully distributed among the networks of the person traveling and sometimes shipped within the United States to close relatives and friends living elsewhere. These exchanges reproduce transnationally the moral and affective ties of sharing through which people become rural Creole (Abranches 2014).[3]

Gathering Together inna States

Perhaps even more critical to replenishing the feeling of being Belizean, of being connected to the heart of Belize, its rurality, and its particular socionatural assemblages, is gathering together with other Belizeans for a meal of Belizean food. Indeed, when a Belizean living in the United States has hicatee, a whole fish, or a leg of wild pig, they often invite relatives and friends over for a day of cooking and eating. A large red snapper served this function at our home in June 2014. Elrick had gone deep-sea fishing for fun off the coast of Texas in the Gulf of Mexico with our neighbor, a white man born and raised in rural Texas.

This Gulf Coast fishing trip affirmed Elrick's adopted Texas hunting and fishing masculinity, a masculinity that is often associated with whiteness (and beer, pickup trucks, and Bass Pro Shops) in media discourse and advertising, despite the popularity of hunting and fishing among every nonwhite racial and ethnic group in the state. But this experience of fishing also became fundamentally rural Belizean as it assembled with Elrick's Afro-Caribbean self. Rural Creole people prepare and eat their fish whole, with head and tail intact, only the innards are removed. Elrick had to catch the attention of the white Texan guides and fish processors who prepare the fish for their customers to take home at the end of the trip, to ensure that they cleaned out the snapper's guts and scraped off the scales, but, critically, without cutting off the head or filleting the fish, as they did for other sport fishermen on the boat. The desire for a whole snapper, the act of "hailing" (getting the attention of) the fish processor, and Elrick's brown skin all came together in the moment. For Elrick this was simultaneously a moment of pride—he caught a large fish that he knows how to process and prepare himself, unlike many of the other sport fishermen on the boat—and a reminder of how he and his foodways mark him as different and other in the dominant social and racial order.

He invited his closest Belizean friends in Central Texas and Houston (almost all of whom identify as Creole) to come to his house to "taste" the fish—sharing the catch among all of them, reconnecting them to their heritage, their roots. He called his sisters in Chicago for advice on the best way to cook it—how he could make it "nice," or make it taste really good. With their advice and his own creative flair, he stuffed the fish with tomatoes, onions, shrimp, and zucchini, seasoned it, wrapped it in foil, and set it to grill outdoors on a barbecue (see figure 7.1).[4]

By the time the guests arrived on this sunny Saturday, the fish was ready to eat—sitting in an enormous foil pan on a table set out on our green grassy lawn. Guests gathered around the fish, half-jokingly, but half-seriously debating who would get the eye and the head—the most delectable parts of the fish. Trudy, a very witty woman I first knew in Crooked Tree in 1990, who had lived in various parts of the United States, and at this moment found herself in Central Texas, asserted, "I only wahn suck that fish eye, ih look nice" (I really want to suck that fish eye, it looks yummy).[5] Another woman visiting from Belize for the week laughed. A third, a woman from the Belmopan area who had been living in the United States in various places, had settled here in Central Texas, and is among our closest friends, loudly and indignantly proclaimed, "Da fi mi di head. Me wan *dat*" (That head is mine, I want *that*). Everyone laughed. The desire to be the friend who got served this delicacy affirmed a sense of coming from rural Belize and the sociality and connection of rural Creole life. But the joking and laughing around this was not only about claims of sociality and

Figure 7.1 Red snapper for a Texas gathering of Belizeans, Georgetown, Texas, 2014. Photo by Melissa Johnson.

which friends are closer but also about how sucking a fish eye or fish head looks to most non-Belizeans. They were well aware that their interest in these parts of the fish marked them as other, amplifying, calling out their racialization and their status as immigrants. These Belizeans, women who most often worked as nannies for wealthy professionals (typically, though not always, white), and men who most commonly work in some part of the construction industry, were well aware of how sucking fish eyes and heads is seen by mainstream Euro/white America as something more suited for a Travel Channel program on eating practices that are "other." Only when they gather among other Belizeans can they indulge their culinary tastes in fish eyes, pig tail, or cow foot soup without being cast as backward, their status as nonwhite other amplified. Here they code-switch into Belizean ways of being on U.S. soil, momentarily evading U.S. projects of racialization (Wilson 2009). Sharing these practices in the safe space of a Belizean American backyard, despite its location in white suburbia and the presence of my complicated Belizean-connected and inflected white self, also serves to further strengthen their sense of themselves as Belizean. The wild-caught fish eye (and only wild-caught fish have eyes worthy of sucking) comes together with gathered Belizeans to generate a socionature that replenishes a sense of what it is to Belizean.

Conjuring Belizean Socionatural Belonging through Storytelling Stateside

Procuring, sharing, and consuming these foods is critical to regenerating Belizeans' sense of themselves as Belizean. Similarly, telling stories in U.S. backyards and kitchens about hunting and fishing trips in Belize also serves to regenerate Belizeanness. This storytelling can happen on the phone, in a video chat, or in person. The conversations (or shared storytelling sessions) that most powerfully replenish are told between individuals who shared the experience but who now may both be living in the United States but thousands of miles apart. Topics vary widely, but discussions of food, and fishing and hunting, are frequent and important. I could share nearly a hundred examples that I have encountered in one way or another. Most commonly, family members and close friends reminisce about events from their past, but these kinds of conversations, sharing experiences in the bush and sharing knowledge about different foods, can also help foster new bonds between immigrants from different racial and ethnic groups.

For example, in 2010, Elrick and I and our children were visited by Ed, a Belizean Garifuna man living in Austin, Texas. He arrived in time to eat the traditional Belizean Sunday dinner we had made: rice and beans that I had cooked and stewed chicken and fried plantain that my husband cooked.[6] The conversation between these two Belizean immigrant men of different ethnicities and upbringings in Belize, who had both lived in United States for many years, was about farming in Belize and the high, dangerous, and wild bush at the Belize-Guatemala border, and black tigahs, and halligata. Each man had countless stories about hunting particular animals in certain places in Belize, travel by foot, jeep, or boat in the bush, and about their farming experiences. The conversation went on for hours: Belize and its particular socionatural assemblages was conjured into being in the kitchen around us. For an hour or two, Central Texas disappeared and Belize filled the room.

Going Back Home

Belizean socionature occupies the imaginations of immigrants living in the United States, but some immigrants are not limited to engaging with Belizean socionature through the occasional hicatee, shared snapper, or hunting tale. Many immigrants are better described as living transnationally, as transmigrants, who physically spend time in both the United States and Belize.

Belizeans living in America return to Belize in a wide variety of ways. Some Belizean immigrant parents living in the United States return to Belize while they are still raising children, with the idea of moving back home. They build a house in Belize and then try to start their own business or find adequate employment to sustain themselves well. Most who follow this path do not last long in Belize because the economic opportunities in the country pale in contrast to the possibilities open to them in the United States. On the other hand, some older Belizeans who raised their families in the United States return to live in Belize after their children are grown. Many of these people spend several months out of every year in each place. Younger migrants living in the United States who "have papers" (either permanent resident/green card or citizenship status) visit Belize as often as they can afford to. And there is a slow but steady stream of U.S.-born Belizean Americans traveling to their home country for short visits to connect with their Belizean heritage, and occasionally as return migrants who settle permanently.

For those rural Belizeans living in the United States who are documented and able to travel, visits home are dreamed about, discussed, and planned far in advance. For Belizeans in the process of obtaining green cards, the prize at the end of the seemingly endless slough of forms, fees, and waiting is the possibility of going home. For those not yet in that process, going home to visit is still imagined, even if it is not likely to happen in the near future.

Featuring centrally in those dreams and plans are the more-than-human elements of Belize: the warm and humid tropical air, the brilliant greens and bright sunshine, and the subtropical plants and animals. The socionature that return visitors find so restorative includes the fishing, hunting, fruit-gathering, horse riding, and fire hearth cooking that enliven their memories of home. Migrants dream about the different fruits (mango, "pia" [a large avocado], mammy, craboo, cashew, and others) that might be in season, whether the water in Crooked Tree Lagoon or Spanish Creek will be low enough for fishing to be fun, or the tropical soils in New River Lagoon Pine Ridge too rain-soaked, thick, and slick for hunting.

Belizeans returning home not only bring the desire to participate in rural Belizean socionature in their home community, they sometimes also bring material wealth and resources that can facilitate this practical engagement—money for gas, boats, fishing gear, camping gear, cooking utensils. Belizeans abroad prefer specific times of year to return home: Easter is the most popular, at least partly because it occurs in the dry season, followed closely by Christmas, when the weather is cool, and September when Belize celebrates itself with national holidays. The dry season is also a particularly good time to go on extended fishing and hunting trips. In 2011, two of Elrick's sisters living in Chicago organized a trip home. Their time in rural Belize included socializing with family and friends, traveling to different relatives' homes and villages, and, of course, lots of fishing and cooking. A highlight of their trip was a day they loaded two boats full of people (a mix of Belizeans visiting from the United States and relatives from their home village), gathered together fishing gear, sun hats, ground food (plantains, cassava, coco, and other tubers and suckers), and

Figure 7.2 Return migrants, author's sisters-in-law on a fishing expedition, Southern Lagoon, 2011. Photo by Elrick Bonner.

set out for a full day's fishing down Spanish Creek, with a boil-up picnic planned for the banks of Southern Lagoon (see figure 7.2). The day was rich with reliving childhood fishing fun and the adventure of making "cyamp out inna bush fi mek byle up" (to set up camp out in the bush to make boil up). A digital camera captured every moment, the photos memorializing rural Belizean socionatural re-engagement; connections with water, mud, sun, crana, bay snook, cassava, coco, and making Belizean food on the tangled banks of Southern Lagoon.

Two other important points about this trip concern two of the travelers. One person on this return visit was Travis, the Chicago-born African American husband of Cindy, one of Elrick's Chicago-dwelling, newly green-carded nieces. Through participating in this trip, he became more entangled with Belize and Belizeans than he had already been by virtue of marrying into the family. Even before this trip, Travis had enjoyed fishing and hunting when he could in the United States, but this trip connected these activities to the bond he had with his wife, his children, and his in-laws. Another unusual enrollee in this assemblage was one of Elrick's sisters, Mary, who has lived in a variety of different places throughout Belize, and who rarely leaves the village center when she is in Lemonal. She was excited to join this trip with her visiting sisters. Through taking this trip with her States dwelling sisters, she also relived childhood memories of fishing and village life, and being a "bush giahl." In this way, return

Figure 7.3 Hunting trip with visiting migrants and current residents of Lemonal, New River Lagoon, 2012. Photo by Elrick Bonner.

migrants intensified the importance of the bush for a rural Creole woman still living in Belize.

Fishing trips are neutrally gendered, but hunting trips that bring return migrants back into relationship with the more than human in Belize are dominated by men (occasionally, a woman who particularly enjoys hunting takes part). In 2012, in the dry season before Easter, which coincides with the height of the nature tourism season, the hunting trip I described in the book's opening story brought together two Belizean men who live primarily in the United States with about eight other men from the village of Lemonal (see figure 7.3). These men gathered together boats, tents, machetes, guns, fishing nets, and other hunting and camping gear to catch fish and turtles and hunt gibnut, wari, and other game meat. The trip was made both more comfortable and successful (in terms of catch) by money and things supplied by Belizeans from "foreign."

The reassembling of Belizeans from States into the socionature of this place during these excursions not only strengthens these visiting Belizeans' senses of being Belizean, as gatherings of Belizeans in the United States do, but also offers a respite from their experiences of racialization and being the immigrant other in the United States. As I argue in earlier chapters, race can assemble differently in Belize, the skin colors, facial features, and hair types that mark blackness come together with skill, knowledge, and sociality. These assemblages weaken the dominant racial structure of white supremacy that is a backdrop of everyday life for Belizeans in the United States. While the contours of color are always present in the interactions of rural Belizeans, and these local politics of color articulate with global racial forms, the rigid racial classification that African-descended Belizeans experience in the United States is temporarily suspended.

The importance of this suspension comes into clear relief when the global racial order becomes salient again. This dry-season hunting trip inadvertently brought together two sets of mobilities, socionatural assemblages, and racialized worlds: black Belizean rural Creole hunters in their ancestral lands and white U.S. ecotourists with the Belizean park warden who served as their guide. Just as the hunters were setting out on foot along the banks of New River Lagoon in the morning, they encountered the tourists in a boat. In this moment, global racial orders that cast these Belizean men, in their camouflage clothing and with their guns, as black, simultaneously cast the American women ecotourists as white. At the same time, socionatural assemblages of rural Creole plenitude collided with socionatural assemblages of ecotourist paradise. Each set of people was out of place in the other's socionatural imaginary. This momentary interruption pulled into relief how, for Belizeans visiting from foreign, being with their fellow Creole relatives and friends back-a-bush alongside New River Lagoon was normally a time when their blackness is not the most salient feature of everyday life. The encounter with white people and the white assemblage of ecotourism brought into sharp relief how these men are cast as black racialized other in the white U.S. gaze, or what Rob Nixon calls "racialized ecologies of looking" (Nixon 2011; see also Joseph 2015). But at the same time, the hunters were also fully cognizant of how in the Belizean bush, their embodied selves associate with their bush knowledge and skill. In addition, the sudden presence of the white ecotourists assembled with the "softness" and relative incompetence of whiteness, particularly as embodied in middle-class tourists from the United States, back-a-bush. Encounters such as these loosen racializing assemblages.

Circulations Otherwise

Rural Creole sharing also re-assembles the socionatural in Belize. One example brings Texas deer hunting and processing techniques, in this case the Texas staple of deer sausage made with jalapeños and cheese, into Belizean foodways. In his first years hunting deer in Central Texas in the early 2000s, Elrick was urged by his rural white Texan hunting buddies to try processing deer meat into sausage. He had a local processor make several pounds of deer sausage for him. Our family in Texas ate some, but we had more than we were likely to consume, so Elrick mailed some to Chicago where friends and relatives loved it and begged for more. Our Chicago-living relatives were reminded of Belizean "hard sausage" (a particular kind of well-loved sausage available in Belize) when they tasted it. Then, on a trip home to Lemonal in 2007, he packed a few of these sausages. They were a huge hit. Elrick's Belizean Creole hunting companions in Lemonal (some cousins and a brother-in-law) particularly enjoyed the meat. The sausage was similar enough to a familiar sausage in Belize to make it palatable, and having been hunted by Elrick the sausage coded Belizeanness. But the deer was hunted in States and then made with Texas-stamped jalapeños and cheese, so it had some States sheen to it. The deer meat's Belizeanness was further strengthened by being placed into these networks of exchange. When our family spent Christmas in Lemonal two years later, we made sure to find Texas jalapeño cheese deer meat sausage to bring. Elrick did not have access to a deer lease that season, but one of his Texas hunting friends gave him a deer to use (thus entangling that man into Belizean socionature). Our two sons, ages ten and thirteen at the time, ate Texas-made deer sausage in Lemonal, with their uncles, cousins, aunts, and extended kin. It brought the family together as rural Belizeans at home in rural Belize. For these two Belizean American, light- and lighter-skinned Caribbean boys, jalapeño deer sausage from Central Texas had become a quintessentially Belizean food, carrying with it associations of crowds of kin back-a-bush, as the evening sky darkens, and jokes, stories, and music fill the air. For my now-adult sons, that sausage still conjures the feeling of rural Belize. And for our kin in Lemonal, eating Texas deer sausage made with jalapeño, reconfigured rural Belizean socionature—Texan deer, peppers, and cheese entering the fray.

Facebook: Making Home Here and There

The electronic and digital communication revolution in the twenty-first century has facilitated linkages between Belizeans in the United States and their relatives and friends in Belize (Clary 2014; Komito 2011; NurMuhammad et al. 2016). When I first visited Belize to conduct ethnographic fieldwork in the early 1990s, there were no cell phones and no social media platforms. Instead, people communicated by telephone to the United States on the shared village landline, or if one belonged to one of the very wealthiest households, possibly on one's own telephone. Paper letters augmented this communication, and migrants living in the United States who could, would visit once a year, but this was still a relatively unusual event in the community.

Today, nearly every adult Belizean has a cell phone and can easily and relatively cheaply call, text, or video chat with relations and friends in the United

States. This increased connectivity means that stories about hunting, fishing, cooking, sharing, and life da bush can be communicated across the diaspora easily and frequently. This increased communication is augmented by Facebook and other social media platforms.[7] The ability to capture images with powerful phone cameras and then instantly post these to large groups of friends and relatives intensifies the significance of the particular meats, fish, and activities in nature in these images to being rural Creole, and even to being from a particular village. People become rural Creole, or someone from, for example, Double Head Cabbage, through liking and sharing these posts. These networks of posting, liking, and sharing communicate the feeling and look of place-based everyday life in Belize to second-generation Belizean youth living in foreign, and vice versa (Kang 2009).

I will share only a few representative examples of the constant stream of these kinds of Facebook posts that document and celebrate the socionatural entanglements of people who trace their origins to Crooked Tree, Lemonal, and other rural Creole villages in the lower Belize River Valley. Posts are made by people living in Belize to celebrate village life, by Belizeans living in the United States either dreaming of home or cooking and eating Belizean foods, and by return visitors documenting their travels and experiences and sharing them here and there.

On a weekday evening in early November 2015, Elrick was sitting next to me at our home in Central Texas, on his cell phone, poring through our joint Facebook account. A new photo popped up, a lush green pasture at sunset, five or six Brahman cows—mostly white, one red, some horses, and a Belizean Creole woman in a denim shirt and pink shorts smiling widely in the middle of the pasture. This was his sister, and the cows belonged to each of them, and the horses belonged to another relative. His sister posted the photo with the caption "country life." Within minutes, other people started posting "the best life," "love it," "gial make we eat one of them cow" (girl, let's eat one of those cows), "Yes, Country life is the best life in the world." The posts were coming from all over—Belize City, Los Angeles, New York City, Chicago, and even one from someone else in the village.

Another woman from Lemonal frequently posts photos of "village life." These include photos of freshly hunted game or fish, a plantashe, a pasture, or food cooking on a fire hearth, to list a few. Each one is captioned in a way that reflects her appreciation for this kind of lifestyle. Her kin and friends from all over the Belize diaspora respond and comment—with appreciation and longing. In 2012, when she was a University student tin Belize, who often returned home, she posted this photograph of an iguana.

Figure 7.4 Iguana. Posted on Facebook, 2012. Photo by Bree Banner.

March 18, 2012
Original poster
"Guana Season." (Iguana Season).[8]
5 people like this. [These likes came from Belize and the United States]

Comments:
Female relative in the United States: "lol . . . i wah some eggs only." (I want some eggs, only, i.e., not the meat, which is also eaten.)

Original poster: "hahahhaha, gyal! 3 ah neh cook up da hse last night! Tr M. eat bout 15 eggs lol." (Hahaha girl, we cooked three of them at the house last night. Teacher M. ate about 15 eggs.)

Female friend in neighboring village: "i only want eggs tooo, 5 enuff fi me . . . lol." (I really really want eggs too, 5 is enough for me . . . lol.)

Original poster: "haaha, egg eating ppl!"

Female relative in Belize City: "Po' guana neh . . ."(The poor iguanas them.)

Female relative in the United States: "lol well i am mad now mommy only eat wah lot . . . lol . . . sorry fi daddy lata . . . omg." (Lol well, I am mad now. Mommy only ate a lot, I feel sorry for daddy, later . . . omg.)

Original poster: "hahahahah,"

Friend in neighboring village: "lol . . . e wa ca stand e own wind." (Lol she won't be able to stand her own gas/farts.)

Male cousin in Belize City: "Mien i should of followed my mind and went home last-night!!" (Man, I should have done what I was thinking of doing and went home [the poster is from Lemonal] last night!)

Original poster: "yeah, we mih di expect fi sih you." (Yeah, we expected to see you.)

Male cousin in Belize City: "I know, the person set mi up, cause i was suppose to go Sunday! them i was in double-head Saturday night, and they asked me if i wanted to go home, mien (Man) i should of."

In the comments on this picture of an iguana, a springtime game treat that is particularly valued for its eggs, Belizeans living in rural villages, Belize City, and the United States all comment about how much they would enjoy eating the iguana and its eggs. One man who lives in Belize City, but had been visiting a neighboring village at the time the iguanas were being cooked in Lemonal, laments his decision not to go to his home village that night. A young woman living in the United States hears of what her mother ate and finds it amusing for the flatulence her mother will experience, but also longs to eat the eggs herself. This post is but one of hundreds of similar examples of Belizeans living in rural Belize celebrating their socionatural entanglements by posting photos on Facebook, and thus tying together the Belizean diaspora worldwide. In this way, these photos, cell phones, and Facebook join with the more than human in the images and the people posting the photos into a continuous re-assembling of Belizean socionature.

Posts depicting Belizean entanglements with the more than human, and in particular with the more than human that ties them back to particular places in Belize, are also made by people living outside of Belize. Belizeans living "da States" routinely post photos of "Sunday dinna" and of their Belizean-style cooking accomplishments. In mid-April 2015, a woman from Lemonal who has lived in Chicago for nearly twenty years posted a series of photos of herself, her sister, and brother-in-law cleaning a "local chicken" (a free-range chicken purchased live and killed by the seller when you buy it) they bought at an outdoor meat market in Chicago. The photos depicted the arduous process of plucking, washing, and properly butchering a large chicken. They needed local chicken so that its meat would be tough enough to "stew" (braise) for long enough to make "steam flour." Steam flour itself is a food for which Lemonal is famous; it is a way of making a large flat dumpling (from flour, water, shortening, and salt, with no leavening agent) that is steamed until it resembles, to my mind, a very wide lasagna noodle. The caption for the series was: "Who say we don't do this in america. (poster's sister) and (poster's brother-in-law) doing a thing for steam flour." The comments included: "We'll be right over" from a Belizean in Los Angeles, and "Looks nice" from someone in Belize City.

Another important category of posting is of meals that include foods brought directly from Belize that one can get nowhere else, such as hicatee from Southern Lagoon or tuba from Spanish Creek. Figure 7.5 is one of a series of photos of tuba, a less often caught river fish from Spanish Creek and Lemonal. The woman posting, Stephie Anthony, known for her wit and way with words, has lived in the United States for a long time, but returns to Belize frequently, and often visits Lemonal where she spent much of her childhood; and being from da bush is very important to her self-identity (which she makes clear through Facebook posts).

Figure 7.5 Fried tuba. Posted on Facebook, 2015. Photos by Stephie Anthony.

May 20, 2015
Stephie Anthony
It's been 30 years since I ate a Tuba . . . thanks to my great cousins and mmmm, tonight I ate two directly from Lemonal . . . alongside some fresh

tortilla and hot shop tea. I couldn't do the wings tho . . . I don't know how to debone them. #villagelife #riverfish

15 people like this. [These likes come equally from Belize and the United States]

Comments
Female relative in Los Angeles: "Lol . . . Fish have wings then."

Stephie Anthony: "Wait . . . I thought the part up by the head was called the wings???? Is it something else?? Omg. Yes . . . the bottom is the tail and then the top part is the head and the two sections with all the bones are the wings. Stop try fi confuse mi. Ah wa clap yo!!" (Stop try to confuse me. I will slap/hit you hard!!)

Stephie Anthony: "I even ate the eyes. I couldn't suck them out tho . . . I picked them out with my fork. They were yummy."

Female relative in Los Angeles: "Eat di bloody fish like yoh know fi eat it and stop di play rass . . . yes dats what we bushy call it wing . . . neva really made much sense to me . . . lol . . ." (Eat the bloody fish like you know to eat it, and stop being an ass; yes, that's what we "bushy" people call it: "wing," never really made much sense to me . . . lol.)

Female relative in New York: "I dnt eat that part eda, u good fi eat fish {w}id bone I cya tek it, look nice y{t}ho." (I don't eat that part either, you're good/strong/admirable to eat fish with bone, I can't take it, it looks good though.)

Stephie Anthony: "Lol I tried to suck them out, but I saw some slimy white stuff, so I had to use the fork instead. I did chew up some of the fins tho . . . they were nice and crunchy. I really really enjoyed them."

Stephie Anthony: "Oh, and I sucked the teeth and chewed that bone. Ahhhhh . . . the pleasures of simple things . . . priceless!!"

Female relative in rural Belize: [tagging the relative in Los Angeles]: "I only eat base snook tail cause I nuh like mess with fish bone." (I only eat bay snook tail because I don't like to mess with fish bones.)

Elrick's sister in Chicago: "Them fish look dry and nice I mi wah crack up them." (Those fish look dry and yummy, I wanted to chew them up.)

The playful dialogue in the comments made by relatives and friends highlight themes I discussed above: how ideas of what is bushy or not (calling fins wings, sucking the eye) are often at play in engaging with Belizean socionature. These ideas are in turn entangled with a complex mix of humor, defiance, a sense of being marked as other (and that marking carrying tones of evaluations based on racialization and cosmopolitanism), and the deep longing Belizeans have for foods from home. In addition, the mix of Belizean Kriol and sophisticated American English contributes to the negotiations of status that take place in these comments.

Through Facebook posts, Belizeans also highlight how they continue becoming rural Creole by hunting and fishing in the United States. In November 2015, I got a text from my husband Elrick, who almost never texts. But he managed to figure out how to send a photograph to me because he really wanted me to post it on Facebook. It was a photo of a freshly killed eight-point buck. He shot it on Coolie Ranch, a place about an hour northwest of Houston, where an extended family of Trinidadians have land for hanging out and hunting.[9] Finding an affordable way to hunt deer is difficult here in Texas, where most good hunting land is privately held and leased out at increasingly exorbitant prices, and Elrick had not hunted in Texas in a few years. He was in hog heaven. He loves to hunt, but this was also a time in which he performs a particular masculinity that he excels at. He is the best hunter among his Belizean and Trinidadian friends in Central Texas, and they admire that about him. Yet, at the same time, these Belizeans who emigrated from cities and towns in Belize also gently make fun of his being bushy. When he hunts successfully, as he had that afternoon, being bushy is primarily a good thing.[10] And he wanted everyone in his social networks, from Lemonal to Crooked Tree, Los Angeles, Chicago, and New York to see this success.

Figure 7.6 Elrick Bonner and his first deer of the season, Coolie Ranch, Texas. Posted on Facebook, 2015. Photo by Elrick Bonner.

November 9, 2015
Melissa Johnson
"Seedys first buck of the season!!!" [Seedy is Elrick's nickname]
77 people like this. [Belizeans and non-Belizean Americans alike, from all over the United States and Belize liked the post; 60 within 5 hours of it being posted]

Comments
Elrick's sister in Chicago: "Aye . . . mi brother!! . . . a see u di do a thing." (Aye, my brother, I see you are doing a thing—Belizean Kriol expression . . . doing something good/industrious.)

Female friend in Crooked Tree: "Bush man fi life." (Bush man for life.)

Female cousin in New York: "Oooh dear meat, nice job cuz." (Ooo, deer meat, nice job, cousin.)

Female cousin in Chicago: "Nice seed make sure you enjoy the liver heart kid with some stream flour." (Nice, Seed! Make sure you enjoy the liver, heart, and kidney with some steam flour.)

Female cousin in Los Angeles: "Wow . . . nice job seedie."

Brother-in-law in Chicago: "Nice one bro lee." (Nice one, brother-in-law.)

Second sister in Chicago: "My precious brother."

Third sister in Chicago: Good job

Cousin in New York: "Nice ☺"

Black American female friend in San Antonio: "Ewwwwwwwwwww!!"

Sister living in Belize City : "Coming your way doc." [Eldrick has multiple nicknames, Seedy is most common, but he is also known as "Doc," short for "Dakta" (Doctor)][11]

Fourth sister in Chicago: "Wow . . ."

White male American retired professor from my department: "That didn't take long!! Nice buck. Congrats."

Female cousin in Belmopan, Belize: "Straight shot Doc. Junie is asking for a back quarter." [Junie is this cousin's brother who lives in L.A.]

Melissa Johnson, tagging female cousin in Belmopoan, Belize: "Seedy says come for the back quarter—but you live too far."

Stephie Anthony, whose post about eating fried fish is included above: "Wait . . . da how Junie get enna dis?? And pan top gat pipple di beg fi ah. [tags mutual cousin living in Belmopan]. Tell Seedy, me da closer cousin . . . da me fi get the back quarter." (Wait, how did Junie get mixed up in this? And on top of that have people begging for him? [tag of cousin] Tell Seedy, I am the closer cousin, I'm supposed to get the back quarter.)

Elrick eagerly awaited the comments that were posted on Facebook. The comments themselves reveal the celebration of bush skills and the importance of sharing these foods. His hunting prowess was admired by people from different villages who live in different places, as well as by white American friends, a black American friend shared disgust that many other friends in the United States have concerning hunted deer. One cousin reminded him of how people from Lemonal should eat deer—steamed flour with the heart, kidney, and liver. And other relatives tried to claim their part of the meat and quarreled with each other over who is more closely connected and therefore most deserving of the meat.

Belizeans living in the United States also use Facebook to share their visits home. Posts from a woman from Crooked Tree who has lived in Chicago for at least fifteen years who visited home in 2012 reveal which elements of Belizean socionature she most cherished. She posted a set of photos of a fire hearth being prepared to fry fish, the fish nearly ready for frying, the result—crana fried hard and nice, and a cashew fruit ripe for the picking (see figure 7.7).

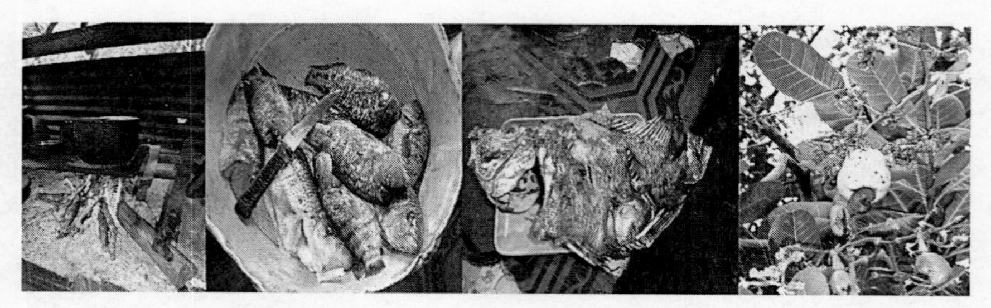

Figure 7.7 Fire hearth, fish, and cashews. A visit home to Crooked Tree. Posted on Facebook, 2012. Photo by Josephine McGee.

There are also second-generation transnational subjects, like a young boy whose mother posted on Facebook a photograph of him holding up a fish he had just caught. His parents, both from Crooked Tree, had been living outside of Chicago, where the boy was born and spent his first years. But his parents relocated to their home village, to the house they had been slowly building as they had money available to send down. They are now settled in Crooked Tree and proudly introducing their son to "life da bush," ensuring that he will be competent, and develop the skills and know-how he needs to be practically engaged in this place. The comments encouraging this young man's fishing abilities came in from Chicago, New York, and Crooked Tree.

Comments
Relative in the United States: "Funk nose ketch one den? Lol." (Funk nose—the boy's nickname, or one of them—caught one, then?)

Original poster—return migrant mom: "Yes gial he got the biggest one." (Yes, girl, he got the biggest one.)

Relative in the United States: "Lol! Sound like fish tea tonight." (Lol, sounds like fish supper tonight.)

Another relative in the United States: "Lol i like that."

Cousin in Chicago: "Dat dah lee man ketch wa pretty lee crana aa!!" (That little man caught a pretty little crana, oh!!)

Yet another relative in the United States: "D boy is a fisherman then." (The boy is a fisherman, then?)

Social media not only allows Belizeans (first-, second-, and third–generation migrants and nonmigrants) to "maintain a degree of presence both 'here' and 'there'" (Clary 2014, 11), but also enables a re-enforcement of place-based attachments to the villages of rural Belize that are the ancestral homes of the Belizeans I describe here (Kang 2009). Indeed, the photos posted on Facebook and the comments those posts garner are themselves part of the continually regenerating socionature that brings rural Creole Belizeans into being. Through these posts, likes, and comments, people become rural Creole in a complex network of relationships between people, plants, animals and places, and transnational living can combine with social media to intensify the importance of particular socionatural configurations to being rural Creole.

Mobilities beyond the Human

Belize has become home to some infamous invasive species: tilapia and lionfish currently attract the most attention, but a host of other plant and animal species have recently extended their range into Belize, including a variety of marine species and Africanized bees.[12] The history of Belize is replete with the introduction of new species, and indeed when Europeans and Africans first landed on these shores, they brought Old World animals and plants with them—both intentionally (for example, cattle) and unintentionally (for example, rats, mosquitoes), generating new socionatures each time.[13] But the story of tilapia is particularly interesting and relevant for rural Creole people.

Tilapia, the name given to a now global food fish, refers to a wide range of different tilapian cichlids. The most commonly encountered of these is known as the Nile River fish and is famed for its invasive capacities. This fish has found a new home in Belize. Indeed, it has arguably become Belizean. The cichlid escaped from some early attempts at tilapia aquaculture in northern Belize in the 1980s. The first escaped tilapia sighting is claimed to be in 1990 in Crooked Tree Lagoon (Esselman, Schmitter-Soto, and Allan 2013). The population grew and dispersed throughout two different watersheds: the lower Belize River Valley and the New River Lagoon river system. Seasonally flooded shallow tropical waters are tilapia's ideal habitat, the seasonal flooding ensuring rich food sources. These waterways that can be so challenging to humans in the rainy season, when the waves on New River Lagoon can reach three to four feet, beckoned tilapia, and the tilapia heeded the call. As the population grew, local fishermen and villagers were concerned. Rumors circulated that these new fish were an extremely successful carnivorous species that preyed on the young of native fish, and so would "finish" (eliminate) all of the beloved river and lagoon species: crana, tuba, bay snook, and others. At this point, few people were interested in trying to cook and eat tilapia, and the few who tried did not think it was "nice" (tasty). In fact, people in Crooked Tree were so upset about the growing numbers of tilapia that during one dry season in the late 1990s, when the water receded so much that it trapped large schools of fish in shallow areas, villagers netted as many tilapia as possible, put them in a big pit, and burned them. The next season, there were even more of these unwanted fish. As villagers began to watch the fish and learn more about them, they realized the fish were more likely to eat algae than baby fish, and concern waned.[14] Over time, rural villagers became accustomed to having tilapia as a fish choice, and are increasingly interested in eating tilapia. The fish began to be popular throughout the country. This shift has occurred slowly and unevenly. As late as 2009, a fisherman in Lemonal pulled in a haul of hundreds of pounds of tilapia, but threw them all back because he would be unable to sell them. In 2013, that same fisherman could as readily sell tilapia as bay snook, and that year filleted tilapia earned the greatest price per pound in sales outside the village. In 2016, a glut of the fish made them harder to sell again. Native, traditional species continue to be more highly valued and desired in rural Creole villages, even as rural Creoles enjoy tilapia. Belizeans' interest in eating tilapia is undoubtedly also shaped by the popularity of tilapia in U.S. supermarkets and the importance of fish to the diets of Belizeans everywhere, including the large migrant communities in the United States. Since people living in rural Belize visit their relatives in the United States, and people living in the United States visit Belize, those circulations likely encouraged the adoption of tilapia as a food fish.

The tilapia that migrated out of aquaculture farms in Belize into the shallow lagoon complex and waterways of Crooked Tree and Lemonal have not remained the same. They have become something else as they assembled with the more than human elements that make up this lagoon complex. Ava Tillett (a leading figure in Crooked Tree) described this change in a news report in 2003, as tilapia were starting to become accepted:

> Ava Tillett, Cook, Crooked Tree Village[15]
> Well to me the way I like it the best is when you filet it or you bake it with vegetables or even hash it. Or you could prepare it and eat it with corn tortilla or flour tortilla.
>
> At first people didn't use to like it. But because of the vegetation that they [the tilapia] are eating now, they change the soft flavouring taste that they used to have; now they have a hard taste. They almost taste like our own local crana.
>
> This fish they come down here in our lagoon, because our lagoon is like a basin from Black Creek and Spanish Creek, so they are right here in our lagoon. (Great Belize Productions Channel 5 News 2003)

Here Ava's enskilment and deep knowledge of Crooked Tree socionature reveals that the new foods these fish are eating is making their flesh firmer than when the fish lived in aquaculture farms eating industrial fish feed. Firm flesh in fish is highly desired among Belizeans, and tilapia that is raised in aquaculture tends to have very soft flesh. But the fish living in Black Creek, Spanish Creek, and Crooked Tree Lagoon, these now free fish, have harder flesh. The new ecologies they are exposed to have altered their embodiment, making these fish appeal to rural Creole tastes.

Another development that indicates how tilapia have swum themselves into Belizean Creole socionature is that shortly after Crooked Tree villagers accepted that tilapia were here to stay (in the early 2000s), an entrepreneur in the community convinced villagers to host a Crooked Tree Tilapia Festival. The festival was held in 2003 and then annually during the dry season, when tilapia catches were the largest. Tilapia has also joined other long-standing festivals, like the agricultural show and cashew festival, through which Crooked Tree has made a name for itself as a location where "village life," as they refer to it in Belize, is celebrated. I was struck by a comment in an online news article describing the 2010 Crooked Tree Cashew Festival:

> For lunch, there were lots of great food, including the usual game meats like gibnut, and deer, and of course, *Crooked Tree's famous fried tilapia* along with stewed beans and white rice, rice and beans, cole slaw, potato salad and fried plantains. You could have washed it down with water or some delicious juices or sodas. (http://www.krembz.com/18th-annual-crooked-tree-cashew-festival/#sthash.aXctAkBJ.dpuf; emphasis added)

In years past, the fish that most represented Crooked Tree to the nation of Belize was crana, but now tilapia competes for that status. Today, lagoons and waterways are indeed filled with tilapia, but native species are still present as well. At least to date, crana, bay snook, and tuba are still common in the catches taken in Spanish Creek, Crooked Tree Lagoon, Poor-Haul Creek, Black Creek, and the ponds along the edge of New River Lagoon. But tilapia have worked themselves in as well, as central features of rural Belizean socionature.

Conclusion

Rural Afro-Caribbean Belizean socionature is assembled transnationally and translocally—through routes of movement and engagement from rural Creole villages to urban areas in Belize and cities and suburbia in the United States. Belizeans who migrate to the United States transform more-than-human items from the United States into rural Belizean socionature (especially wild-caught game and fish), and share these foods and products with kin in social networks spread across the United States. Belizeans living in the United States also gather together to enjoy these kinds of food products and to share stories of hunting, fishing, and rural life. These gatherings offer a place to enjoy being Belizean and eating Belizean foods in a way that temporarily envelops Belizeans within Belizean networks, thinning the racializing structures of their experiences in the United States otherwise.

Many Belizeans living outside of Belize return and replenish their sense of themselves as rural Creole through engaging socionaturally in the villages they come from. They may simply eat fruits, fish, and game meat, and share in networks of exchange, or they may embark on fishing and hunting expeditions. These expeditions can serve to strengthen the importance of socionatural engagement not only to these migrants but also to villagers who remained in Belize, who are enticed to join in on the trips. These trips also provide a way for Belizeans to nourish their sense of themselves in the world, to take a respite from the global racial order—a respite made sharply visible when disrupted by the surprise presence of U.S.-based white ecotourists. Visiting Belizeans also expand socionature in Belize by bringing in new engagements with the world, for example, hunted Texas deer meat turned into Texas-style sausage.

Another realm in which Belizeans transnationally generate socionature is through social media. Posting photos of rural Belizean "village," "bush," or "country life" garners likes and comments from across the diaspora—reconnecting far-flung Belizeans around the idea of life in rural Belize. In this way, Facebook photographs and likes assemble with Belizean Creole socionature. Return visitors can share their re-intensified engagement with village life, and connect people still living in rural Belize to migrants living from Los Angeles to Minneapolis. Social media like Facebook intensify the importance of rural Creole socionature to being rural Creole.

Finally, the mobilities that assemble rural Creole socionature transnationally are not limited to human movement, the more than human can assume some agency in this process as well. Although brought to Belize initially by humans for aquaculture, tilapia escaped farms during flood seasons and figured out how to thrive in the watery ecosystems that rural Creole people call home. Thus transformed into firmer flesh, they then worked themselves into the hearts and socionature of rural Belizeans.

All these examples show that socionature is not simply tied to one place, but can link places together. Belizeans live transnationally through socionatural engagements and these assemble in different ways with processes of racialization. It is through these engagements that Belizeans living in multiple places continue to create economies and ecologies *otherwise*.

CHAPTER 8

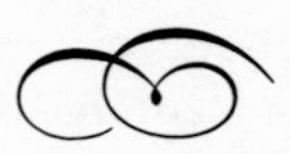

Conclusion

LIVITY AND (HUMAN) BEING

I was putting together the final edits and figures for this book manuscript in September, the month of national celebration in Belize, containing holidays that celebrate all Belizeans to some extent, but still disproportionately celebrate Belizean Creole culture. This is a good time to be home if you are a Creole Belizean living abroad. Elrick had not been home for a September celebration in years and chose this year to visit for Independence Day. Days before he left Texas, he was talking with a sister who lives outside of Belize City and who has visited us in Texas numerous times. Elrick shared with her that he had been "feeling fi some Creole bread" (a yeast bread made with coconut milk) and then they started laughing. I remembered that my sister-in-law really likes the Creole bread I make (which I do in a lazy way, partly with a mixer) and likes it best a couple of days after it is made when it is just a little bit more dry and firm. I piped up and offered to make some and send it for her with Elrick (and promised him that I would save a loaf for us to eat before he went). I felt proud that my Creole bread is good enough to share, recognizing that shared food is especially good food. I am also surprised at myself—how have I not thought to do this before? Elrick and I are often the recipients of Belizean treats sent to us by relatives, and I have enjoyed these treats every time we have received them. I could have, should have, made bread to send so many times, or brought it with me on so many occasions when I have visited. I am still learning. I made and sent the bread, and helped to transnationally assemble rural Creole culture.

A week or so before Elrick was to arrive in Belize, his sister who lives in Lemonal went to clean out the house in Lemonal village that he (and I and our children) usually stay in. The raised wooden house had initially been built by Papa, the family's recently deceased father to whom this book is dedicated. Elrick and I added rooms to the house in the mid-2000s, and it has served as both a place to stay and a gathering point for extended family who live elsewhere since then. It had not been used for about three months, and in that short period of time had been taken over by bats, their powdery and smelly guano a threat to human health. His sister got her young adult son to beat the bats out of the house before Elrick was to arrive, and other family members and friends helped clean the house to make it habitable. In the northern Belizean lowlands, bats thrive and take over places humans have made—knowing how to remove bats from houses is crucial. This is no easy feat, bats can flatten themselves almost to invisibility and slip into the tiniest of openings in a house's siding. But Elrick's nephew knew just what to do and the house was cleaned out swiftly and fully and was pleasant and clean when Elrick arrived. The tangible presence of agentive more than human, and the know-how and skill to live with, among, and against these entities is always a part of becoming rural Creole.

When I talked to Elrick midweek, he had bought the beaverboard needed to seal the house to keep it bat free, and he and his sister were planning to cook a big pot of chicken and have friends and family help fix up the house on the weekend. But when we talked again on Sunday those plans had been set aside. The materials were there to build the house, but people did not feel like doing that work on Saturday. It was September, after all, and the house was habitable. Instead, friends and family gathered to eat, play "block stick" (dominoes) and enjoy themselves and each other's company. And this continued into Sunday, when more family from Belize City pulled in to celebrate the birthday of my "bra-lee" (brother-in-law), Jack. Elrick said Jack had "gone to look for a deer"—the rumor was that someone had had a successful hunt the night before. But Jack's family was also already cooking up a storm, "Sunday dinna"—stewed chicken, rice and beans, potato salad. Elrick teased me, asked if I could hear the chicken sizzling, telling me what the potato salad looked like, and laughing as he told me I would have to go get a fish sandwich from McDonald's—a poor substitute for food. He told me he was playing block stick, and I could hear everyone laughing and having fun in the background. They were outside, a few lights shining in the gloaming, eating, playing games, laughing; brownness and blackness assembling with sociality, love, fun, food, cooling breezes and the bright smells of tropical foliage. This was Sunday in Lemonal, or Crooked Tree, or Bermudian Landing. This was a rural Creole Sunday evening. And as I finished writing this story, my sister-in-law in Chicago posted on Facebook a photo of the Sunday dinna she had just made—beans and rice, stewed pork (it looks like), potato salad, and tomato slices . . . with an orange Fanta. Transnationally living Belizean Creole people re-create village Sundays in Chicago as well as they can.

Rural Belizean everyday lived experience generates an expanded human being. The thinking that dominates scholarly production and popular culture in the Global North and increasingly all over the world has overrepresented a particular conception of the human: a conception based primarily in the experience of white Christian, heterosexual, able-bodied male thinkers from European capitalist societies. Sylvia Wynter conveniently calls this overrepresented figure "Man" (Wynter 2003; Wynter and McKittrick 2015). Man makes it hard for anyone living in the Global North and especially those trained in the Euroversity to see different genres of being human/human being. Man is understood as a separate biological category, a species that is uniquely capable of generating change, that with culture and language sits apart from, and above, the natural world, not as a being emerging out of relation and connection. Man is overwhelmingly white, and Man's (white) existence necessitates the exclusion and subjugation of black, brown, yellow, and red others. Moreover, Man is individuated (not relational), rational (not emotional), and is driven to accumulate (maximize economic gain). Wynter argues that *black experience* in the New World offers a critical source of knowledge to help make visible both the limitations of Man and other possible genres of the human (Kamugisha 2016).

A black-centric, beyond-Man genre of human being is reflected in the increasingly widely used Rastafarian concept of "livity." This word shows up in everyday speech throughout the English-speaking Caribbean and also appears in scholarly work in Caribbean and Black Studies. Livity refers to "way of life" and often includes the idea of *freedom*. In his treatise on freedom in Caribbean societies, Neil Roberts describes livity as "a phenomenology . . . a lexicon, a style of attire, a connection to the environment and the *ital* (natural), a manner of interaction with self, sovereign, and other humans, and a disposition grounded in mutual love and respect," and an orientation away from capitalism (Roberts 2015, 176; Roberts 2014, 190). The love and respect of self and other in this blackened *ital* way of living has the capacity to "constructively change the psychological and metaphysical dispositions of humans and all forms of life." (Roberts 2015, 176).

The natural living and reciprocal relating captured in this term, its imagined alterity to Babylon (by which is meant white supremacist capitalist culture), and its origin in Afro-Caribbean culture are suggestive. The Crooked Tree fisherman whose interview opens chapter 5 talked about his and his fellow villagers' livity. He uses the term to refer to fishing and how that was handed down through generations and how it was what he and his fellow villagers knew. But he is also appealing to a broader sense of the word.

With the ethnographic stories shared in this book, I show how rural Belizean Creole people generate livity. Afro-Caribbean Belizean Creole communities came into being in the violence of the slave trade, slavery, and extractive capitalism. In the very belly of this beast, I argue, these people created alternative subjectivities, ways of being human beyond Man. Rural Creole people continually re-create alternative modes of human being, subjectivities that provide parts of a road map for how to thrive as humans in this era of planetary crisis.

Ecologies, economies, and subjectivities that expand beyond Man have persisted in rural Belize since its earliest days. Renegade European-ancestried buccaneers and pirates in the margins of state power created noncapitalist economic activities as they camped along the shores of British Honduras. As these men, and over time, women and men of African ancestry, settled into these water-soaked lands, they created their own cultural norms and practices. The ecological and economic forms that emerged were tied to, but not determined by, capitalist economic relations connecting Belize to other parts of the world. Figuring centrally into this emerging Creole culture was a practical engagement with the more than human, and networks of sharing centered on plants and animals of this place.

Belize's earliest moments were characterized by a relatively open and fluid racialized political economy. Logwood camps gave way to more entrenched and unequal racial and economic structures as mahogany dominated the economy and slavery and its violence expanded. The northern lowlands focused on in this book became the home not of wealthy whites who lived on coastal islands, but of a polity with a different sensibility. This place was settled by poorer whites, who sometimes owned a few enslaved blacks, "free-colored," who also sometimes owned enslaved blacks, and "free-black" people. Here this multi-hued, multi-embodied group of people generated a variety of ways of sustaining themselves from the tropical lowlands where they lived. With emancipation, colonial discourse and policy encouraged poorer Belizean Creole people to remain living in these places where the land was not as rich for agricultural development. This ensured that they would be available for wage-labor work in the timber industry. But rural Creole people nonetheless crafted ways to live that were not dependent on full-time, steady work in the timber industry. They farmed, hunted, fished, engaged in "kech and kill" and took advantage of whatever opportunities came their way, but, critically, on their time and in their own ways.

Rural Creole people live in landscapes full of powerful, agentive entities, from water that causes floods to unpredictable waves that can swamp dories, deadly halligatas (crocodiles) and tommy goffs (fer-de-lance), and the elusive Tata Duende who can lead one astray. Rural Creole becoming entails perennial struggle with keeping the bush cleared from yards and pastures. It requires having a full sense of the dangers posed by the more than human, and the agency and canniness of the entities with which Creole people share this place.

Rural Creole people become who they are through deep entanglement with the more than human. Their extensive knowledge and enskilment in this place convene with fish (from crana to tilapia), hicatee (turtle), cattle, and cashews, among a host of other more-than-human things. Becoming Creole involves, for example, knowing how to move through this swampy landscape to find the best spot to fish, what fishing technology is needed, how to prepare and cook the fish, and knowing the pleasure of eating the fish. Belizean Creole enmeshment with the more than human is also symbolic. Storytelling, musical production, and Kriol linguistic play, all so important to everyday life in the Afro-Caribbean, are centered on the animals, plants, waterways, and more that make up the lowlands of Northern Belize. In all these ways, Creole people generate a human being that emerges out of relation and entanglement and that recognizes the significant power and agency of the more than human.

Rural Creole becomings assemble blackness, brownness, and whiteness with the more than human. In moments of casting the bush as "backwaad" and more black, and in the privileging of clear or light skin over dark, these assemblages can conform to Man's white racial frame. But more commonly, different Creole embodiments, brown skin and black skin, "kun-ku" (short) or sharp noses, "seedy-headed" (very kinky) to straight hair, assemble with enskilment, knowledge, intelligence, linguistic prowess, musical fun, creativity, sharing, belonging, love and respect. In this way, race "proliferates" (Saldanha 2006) beyond the structure of the white racial frame, generating liberatory brownness and blackness.

Rural Creole livity is also oriented toward sharing and sociality, a sharing and sociality that is even more accentuated when game meats, fish, and other natural things from this place circulate. Sharing is a key part of rural Creole culture and a measure of one's goodness as a person: sharing the bounty from a hunt or a fishing trip is expected, and some of the best hunters have little interest in selling the game they catch, preferring to share with kin and family. People also share labor in social gatherings to clear land, build fences, or construct houses. As work, food, and sociality all convene together, people participate as much for enjoyment as for anything else.

The alternatives ecologies, racializing assemblages, and economies that emerge out of Belizean socionatural arrangements in this part of the world have persisted through the shifts the area has experienced. Rural Creole people have lived alongside biodiversity conservation projects, hosted a growing ecotourism industry, and are incorporated into circuits of transnational living created by emigration. While it might be expected that legal limits on the use of wild products, the commodification of nature under tourism and the physical separation of rural Creole people from the places where they were raised would attenuate socionatural connection and the noncapitalist ways of living they have forged, and would re-inscribe white racial frames, the story is far more complex. In interesting ways, each of these developments has served to intensify some aspects of pre-existing assemblages of rural Creole socionature. For example, conservation has led to more halligata and a sense of human fragility, and limits on hicatee have intensified its importance as a food that makes one Creole. With tourism rural Creole people have become scientific experts and top chefs, re-assembling excellence with blackness. Sharing game meat and fish across transnational borders as well as images of the more than human on social media brings people into being as Belizean Creole whether in Chicago or Crooked Tree.

In sum, rural Belizeans are (and have been) entangled with processes of racialization and capitalist economies of timber extraction, conservation, and tourism, but at the same time they generate economies, ecologies, and racial assemblages from their creative agency, and these have existed alongside enslavement and racial capitalism all along. In these efforts, rural Creole Belizeans expand human being.

Maybe what is needed in this moment of planetary crisis and growing inequality is just this kind of expanded human being. Cultivating livity, ways of living and being in reciprocity and love, might encourage "kin-making"—a way to "practice better care of kinds-of-assemblages (not species one at a time)," to take care of all "earthlings" as Donna Haraway exhorts us to do (2015, 162) Rural Creole livity arose despite the most violent and marginalizing histories of racialized capitalism. Do rural Belizean Creole modes of human being offer healing orientations toward this planet earth, and toward love, for human becomings in different places, and with different histories?

At a party in my Central Texas backyard, I talk about this book I am working on with my white working- and lower-middle-class neighbors, several immigrant Belizeans, and others. I describe my contention that people become who they are through their relations with the more than human, whether it is the U.S. teen and a cell phone or a Belizean fisherman and a tarpon, and how important sharing and creating community through sharing is to the Belizeans I know. My thirty-four-year-old neighbor and friend, born and raised in South Texas, working at this point as a heavy equipment operator for a local municipality, nods his head vigorously. For every example I share from Lemonal and Crooked Tree, he has a matching one from growing up in South Texas, or from living right here on the outskirts of Austin, and creating kin with neighbors and friends and fish and deer. A few days later, another white American friend in his thirties brings us a pail of fresh tilapia he "spiked" (speared with a bow) somewhere not too far from here. My black husband is thrilled, and finds space in the freezer. In these moments, our livity expands. The fish are whole, Elrick's skill and love assemble with his embodiment, with the bucket, with the tilapia, with our friend's whiteness and rural Texas culture, and all of this comes together with love and community. This was a moment of livity, of relations of reciprocity and love between people and the more than human, and racializing assemblages that move against hegemonic orders. Alternative ways of human being exist in pockets everywhere. It is these moments that should be identified, supported, and expanded as we try to find ways to live in the Anthropocene.

Appendix: Kriol Words and Phrases Used in Text

Spelling for Kriol has not been fully standardized. I am using spellings that I have seen; Decker 2005 has some of these words spelled differently in Kriol.

GENERAL

backwaad	Not modern
bakra	White person (perhaps of Igbo origin, see Cassidy and LePage 2003, Decker 2005)
blockstick	Dominoes
bra-lee	Brother-in-law
broad Kriol	The version of Kriol that contains the most West African elements
bushy	Of the bushy, wild, backward (see chapter 3)
bwai	Boy
clear (as in skin)	Light-skinned
cold seed	Goosebumps
coolie	Commonly used, but sometimes derogatory description of people of East Indian ancestry
cranky (as in dory)	Wobbly, difficult to control, easy to capsize
cuss out/cussing	Cursing, or public criticism and ridicule. When done correctly this criticism ridicules for things that are accurate about the person
dirt hole	Garbage pit
dory	Canoe, usually made from one large tree trunk
facey	In your face, cocky, presumptuous
finish	To become empty, to no longer be there. "Ih finish" can mean there is nothing left
fool-fool	Especially stupid (any adjectives can be repeated in this way for emphasis, as this one often is)
foreign	From anywhere other than Belize, but especially the United States

giahl	Girl
good hair	Straight fine hair (as opposed to very curly, kinky, coarse hair)
haad	Tough, able to deal with the rough patches; also, food that is not soft and ice
hall	The main shared living area in a house
hail/hailing	To get the attention of, to call out
idle	Useless, up to no good, causing trouble
ih	Gender neutral pronoun, "he or she"—Kriol does not distinguish
kech and kill	A way of making money by doing whatever one can for short periods
know-well	Know-it-all
Kriol	The lingua franca of Belize, and the language spoken by Belizean Creole people
kroffy/Kruffy	Bushy
kunku	Short, chopped off. Often refers to noses.
lee	Little
livity	Originated in Rastafarianism; means way of life and has other associations (see chapter 8)
modahn	Modern, not backward
new neega	New nigger—see chapter 3. Refers to an African slave new to the Americas who does not know how things work
nice	The word most often used to say that food is delicious. Degree of deliciousness is distinguished by amount of emphasis on the word
off the main	Far from main roads, waterways, cities
old head	Elder in the community, who has accumulated wisdom
papers (papers coming up, having papers)	Immigration papers, especially in the United States documented status with a pathway to the green card or permanent resident status
peel	Bare, like a peeled fruit, often refers to balding
peg	Stick, can also be a spear for fishing
pint bottle	Glass bottle (such as the kind that holds beer or soda)
pit-pan	A skiff
poppy show	A fool-fool performance, or way of behaving; silly, stupid
prize	Value, desire
quarrel	Engage in cussing, see above
raw Kriol	The version of Kriol that contains the most West African elements
seedy, seedy-headed	Hair that is very kinky and rolls up into little balls (like seeds)
shame/shamed	An important emotional state in Creole culture, to be avoided at all costs; to be caught in an unseemly light; some people have shame around eating in public. A particularly sharp version of feeling embarrassed

shape/no shape	Referring to the shape of a body, especially women, but also men. No shape means having a very flat bottom; and for women especially having no discernible waist
skahn	scorn
soff	Soft; unable to handle the obstacles that arise in life, easily giving in to the desires of others; well-cooked food is soff
speaking	To talk in English
States	The United States
sweet	To please someone, make someone feel good; pleasing, feel-good making
talk fool	To playfully and adeptly use language, storytelling, and joking as entertainment
village life	The celebration of rural living that is seen as authentically Belizean
waika	From the Miskito coast, implying some indigenous dimension, and connection to dangerous spirituality
wile	Wild, shy, in the way that a deer is afraid of humans
willing	Willing to do whatever is asked, eager to please, ready to help
wite giahl	White girl
wussus (of Kriol)	The worst Kriol, the most African version

GENERAL PHRASES

"bocatora language"	Mixed up, slow language (named after a slow turtle), a derogatory way to refer to Belizean Kriol
"Da sin"	It's a sin. Common phrase
"Di rite ya di do notteen"	I'm right here doing nothing. The ideal state of being—to have nothing that needs to be done, to have time to socialize
"Di rite ya di fite life"	I'm right here working hard to get by
"doing a thing"	Doing something interesting and/or particularly useful
"dumb to di world"	Unsophisticated, bumpkin
"ghetto youth"	Poor young people (black)
"handle one's candle"	Take care of oneself, be capable of coping with what might arise
"maka parrat"	To mimic someone (mock), but playing on the homonym of macaw parrot
"play rass"	Behave obstructively
"travels with"	Has chronically (especially used to describe medical conditions, pain)

PLANTS AND PLANT-DERIVED ITEMS REFERRED TO IN TEXT (SCIENTIFIC NAMES FROM BELETSKY 1999, BRIDGEWATER 2012, GILLETT AND MYVETTE 2008, MACKLER AND SALAS 1994)

bitters	Mixes of herbs, leaves from shrubs, tree barks that have a bitter flavor (there are many possible different mixtures, each one designed for particular medicinal effects)

breadfruit	Popular fruit planted in most villages, most often eaten fried. Native to the Pacific Islands
bullet tree	A tall tree with extremely hard wood. *Bucida buceras*
bush medicine	Includes bitters and also other herbal and plant-based medicinal products
calabash tree	A tree with spines and small leaves that produces large gourds. Used as a marker to identify places in rural Creole landscapes. *Crescentia cujete*
calaloo	A spinach-like leafy vegetable used widely across the Caribbean
cassava	A tuber, "ground food," that is critical to rural Creole foodways (along with yams and coco)
cashew	The only fruit that grows its nut externally; common in Crooked Tree. The fruits and nuts are important sources of income; an annual festival is organized to celebrate it, and it is identified with Crooked Tree. *Anacardium occidentale*
cedar	Important and common tree. The lightweight but strong wood is good for dories as well as bowls and utensils for the kitchen *Cedrela mexicana*
chicle /chiclero	The sap of the sapodilla tree that can be boiled down into gum. The original material that chewing gum was made of. A chiclero is a person who gathers chicle for a living
cohune palm	A marker of rural Belizean landscapes with its vase-like growth. The nuts can generate a delicious oil. Different rural Creole people make this oil to sell. The palm marks good, rich soil for making plantash. *Orbigyna cohune*
cocoyam (coco)	An important tuber, or "ground food." Native to the Americas. Important ingredient of boil ups, also used in other ways. *Xanthosoma sagittifolium*
craboo	A very popular seasonal fruit. It lasts a long time when put in bottles with water and sealed. Popularly eaten with sweetened condensed milk. *Byrsonima crassifolia*
ginnep	Also kinep. A popular seasonal fruit (my favorite!), widespread through the Caribbean. *Melicoccus bijugatus*
juke-mi-back	A prickly bush, the Kriol phrase translates as "poke my back"
kiss-kiss	Wooden tongs made from a long piece of pokenoboy palm that is heated over hot coal and then bent and held in place until it retains the bent shape. Made to use for removing coals and other hot items
logwood	The dyewood, and now good fence-post wood, that honed European interest onto this part of the world. *Haematoxylum campechianum*
mahogany	The furniture and railcar wood of the nineteenth century that is now also a popular furniture wood. Made some British Hondurans very wealthy. *Swietenia macrophylla*
mammy apple	A native fruit tree, the fruit is popular. *Mammea americana*

mango	One of the delights of rural Belizean yards. *Mangofera indica* (and possible other species and or subspecies)
my lady	Tree cut for timber. *Aspidosperma megalocarpon*
old lady stinking toe	Possibly apocryphal, but perhaps a shrub that has a stinky flower
pia	Avocado. *Persea americana*. In rural Belize these are large, light green, and smooth-skinned and taste delicious
pimenta palm	A palm tree with slender trunks. The wood is strong and lasts a long time. Used for housing, fence posts, and other purposes. *Paurotis wrightii*
pine	The landmark tree of the pine savannah. Cut for timber. *Pinus caribaea*
ping-wing	A common prickly plant that makes traveling through the bush challenging. *Bromelia pinguin*
plum	A variety of plums (May plum, August plum, golden plum, governor plum). These are all cultivated and are medium-size fruits with a relatively large center nut
pokenoboy	Also pork and dough boy. A spiny palm with delicious fruits. Also used to make kiss-kiss (*Bactris major*). "Kis-kis" is a Moskito word (from the Nicaraguan coast)
Santa Maria	An important timber tree. *Calophyllum brasiliense*.
sapodilla tree	Sapodilly. The tree that generates chicle. Also has a delicious fruit. *Manilkara zapota*
suckers	Plantain, bananas, and other plants that grow in similar ways
tubroos	Large spreading tree that characterizes high ridge. Also Guanacaste. *Enterolobium cyclocarpum*
yam	Large tuber that comes in yellow and white forms; important ground food. *Dioscorea sp*

REPTILES, SPIDERS, MAMMALS, BIRDS, AND FISH (SCIENTIFIC NAMES FROM BELETSKY 1999, BRIDGEWATER 2012, AND MACKLER AND SALAS 1994)

Reptiles

bocatora	Wayumu, common slider, sun turtle, ornate terrapin	*Trachemys scripta*
halligata	Morelet's crocodile	*Crocodylus moreletii*
halligata, saltwater	American crocodile	*Crocodylus acutus*
hicatee	Central American river turtle	*Dermatemys mawii*
iguana, guana, bamboo chicken	Green Iguana	*Iguana iguana*
laagrahed turtle	Sambodanga, tortuga lagarto, snapping turtle	*Chelydra serpentina*
Tommy goff	Fer-de-lance, Central American lancehead	*Bothrops asper*

Spider

tri-antelope	Mexican red-rumped tarantula	*Brachypelma vagans*

Mammal

baboon	Black Howler Monkey	*Alouatta pigra*
gibnut	Paca, tepezquintle	*Cuniculus paca*
hamadilly	Armadillo (nine-banded)	*Dasypus novemcinctus*
Indian rabbit	Bush rabbit, Central American agouti	*Dasyprocta punctata*
mountain cow	Baird's tapir	*Tapirus bairdii*
peccary	Collared peccary	*Pecari tajacu*
quash	Coati, pizote	*Nasua narica*
tigah cat	Includes both ocelot *Leopardus pardalis* and margay	*Leopardus wiedii*
tigah (leppahd)	Jaguar	*Panthera onca*
tigah, black	Jaguar (black)	*Panthera onca*, black variation of Jaguar
tigah, red	Puma, mountain lion	*Panthera concolor*
wari	White-lipped peccary	*Tayassu pecari*

Birds

currasow	Great currasow	*Crax rubra*
turk	Jabiru, filllymingo	*Jabiru mycteria*
quam	Crested guan	*Penelope purpurascens*
toby full pot	Great blue heron	*Ardea herodias*
whistling duck	Black-bellied whistling duck	*Dendrocygna autumnalis*

Fish

baca	Blue catfish	*Ictalurus furcatus*
bay snook		*Petenia splendida*
crana		*Cichlisoma uropthalmus*
tarpon	Tarpon	*Megalops atlantacus*
tilapia	Tilapia	*Tilapia mossambicus*
tuba		*Cichlisoma friedrichsthali*

OTHER WORLDLY CREATURES

ashishi pompi	Small gray creatures the color of soot that become lively as a fire dies down in a hearth
devilments	Any of the supernatural entities of the rural Belizean landscape
jack-o-lantern	An orb of light that forms over the savannah; it can change in size and lead one astray if one follows it
old heg	A ghostly female figure

Tata Duende	A creature of the forest. A small man with backward-facing feet, no thumbs, and a large hat that is known particularly to lead children astray
wari massa	A giant wari who protects the forest and the creatures in it

LANDFORMS, LANDSCAPE DESCRIPTORS USED IN TEXT

back-a-bush	Far from any city or village
bank(s)	A spot on a creek, river, stream or waterway edge that is lived on by someone, or claimed by someone
barcadere	A good landing spot on a waterway
bush	The uninhabited "wild" parts of a landscape (see chapter 3)
bush, high	High and dense tropical or subtropical forest
caye or cay	In inland rural Creole places, a concentration, or copse, of tall and dense forest. In marine areas, a small piece of land, mangroves and sand
cohune mud	The rich deep black mud that accompanies cohune ridge
dump	A causeway
Indian hill	Maya sites. They are usually raised above surrounding landscape
man drop rain	Very large heavy raindrops
off-the-main	Places that are out-of-the way, difficult to access
picado	Usually "lee picado road." Small trails or roads cut through dense bush
plantashe	Agricultural lands for cultivating. Making plantashe means chopping bush to prepare the land, planting, tending, and reaping cultivated plots. Items cultivated range from ground foods to suckers to rice, corn, beans, tomatoes, papers, and so on
ridge, broken	Less dense tropical forest
ridge, cohune	Cohune palm forest, dominated by cohune palm, dense tropical vegetation; has rich deep mud
ridge, high	Broadleaf forest tropical and subtropical forest. Dense and tall, dominated by mahogany
ridge, pine	Pine savannah or pine forest/savannah, stretches of savannah (grass and reed lands) punctuated with pine (*Pinus caribaea*) and a few other types of trees and containing occasional cays of high ridge
the dry	The dry season(s). The primary dry season occurs in April and May; another small dry season has historically occurred in August. All of this is changing with global climate change
tigah camp	A hunting camp for rich white men who were guaranteed a jaguar kill. Jaguars were captured by rural Creole hunters, then caged until the white Americans came for their hunt, when the jaguars would be released and chased by the hunters
To the back	Toward Guatemala and away from human settlement
walk, coconut	An orchard of coconut trees

walk, lime	An orchard of lime trees (Lemonal got its name from this: lime walk/lemon walk/ Lemon-al)
works	The location of logwood and mahogany cutting businesses owned by one man in Belize's early history

FOODS AND FOOD-RELATED TERMS

black tea	Tea, the drink. Typically mixed with condensed milk and sugar
boil up	A Belizean dish made by boiling together ground food, suckers (plantain or unripe banana), sometimes eggs, sometimes fish, sometimes pig tail. Served with coconut or cohune "fat," or oil, and a tomato and onion sauce
cow foot soup	A Belizean dish made with the lower legs of a cow/beef, ground food, and vegetables. The sinews in the cow foot give the soup a thick, goopy consistency. It is good for hangovers, builds strength, and male virility
cuttobrute	A candy or sweet treat made with shredded coconut, butter, and sugar
dinna	The main meal of the day. Historically eaten at midday, and this remains the case for many still living in rural Belize. Rarely made without rice. When no rice is served, corn is
fat	Refers to oils, lards, any kind of fatty substance used in cooking, such as coconut fat
fiyah haat	Fire hearth, outdoors; the way everyone cooked before the late twentieth century. Still used for cooking in large pots for big gatherings, or to create the particular flavor that only a fire hearth can create, or for frying foods to keep odors from getting caught inside a house
fried jack	A fried bread served for breakfast and supper. Also "fried cake."
fried plantain	A staple of Belizean food. They need to be soft and ripe, and then are fried until the sugar in them carmelizes and browns the edges
ground food	Tuber foods—yams, cocoyams, cassava, potatoes. Staple items of the rural Belize diet
hard sausage	A cured dense sausage, similar to summer sausage in the United States.
johnny cake	A biscuit made with flour, shortening, salt, water, and baking powder. They last a long time. The name may come from "journey cake." They are often taken on multi-night trips
local chicken	What might be called free-range chicken, raised in the village and not in pens, or a factory farm. The meat has a particular flavor and takes a longer time to cook until it is tender
pig tail	Cured pig tail (in brine) is another Belizean staple. Tail tips and other sections are put in boil ups, and used to flavor a

	pot of (red kidney) beans. The tails themselves are a much desired delicacy
rice and beans	Rice cooked with beans (red kidney beans). The beans are first softened and then the rice and beans cooked together with coconut milk. This is both a Belizean staple and a fancier way to cook rice. Rice and beans are served at celebrations, funerals, weddings, and on Sundays
roast and broke cashew	Cashew seeds grow in a large poisonous husk. These are charred on an outdoor fire (roasted), so that the husks can then be broken off
sere	Maybe also spelled "cere"; a fish soup with onion, coconut milk, and gently boiled fish (complete with bones)
steam flour	A dumpling or noodle-like concoction made with flour, shortening, salt, and water, rolled thin, and steamed, ideally over a pot of seasoned local chicken. Lemonal is known for this food item
stewed beans and rice	White rice with coconut milk served with a spoonful of softened and seasoned red kidney beans on the side (beans may also have pig tail, or dumplings or salt beef in them).
stewing/stewed chicken	The standard way to cook meat in rural Belize is to fry it in coconut oil until it is dark brown, and then to "stew" or braise in just enough liquid until it softens
Sunday dinna	On Sundays, the standard midday meal is rice and beans, stewed chicken, potato salad (Belizean style), and fried plantain
taste	People will say they want to "taste" your food, to see what your cooking is like, or because they are interested in knowing how a food tastes. This is not about eating to sate hunger, but rather about sharing different foods and cooking styles
tea	The morning and evening meals, and also the drinks that accompany those meals. The drink can be Milo (a malted chocolate drink mix from Britain) or sweetened condensed milk and water, or black tea with milk. The food is typically something made from flour (johnny cake, flour tortilla, fried jack) and some beans or meat from earlier in the day or the night before, or cheese and avocado with coconut oil, or egg
wangla fudge	Sesame seed fudge

Notes

CHAPTER 1 — INTRODUCTION

1. Whether to spell this language as "Creole" or "Kriol" involves some complicated politics. I have chosen to use "Kriol" to distinguish when I am talking about the language from when I am talking about the people.

2. It is not insignificant that this encounter occurred about one year after Trayvon Martin was killed. The white women ecotourists were most likely from the United States and would have been exposed to the extensive coverage of this killing of a black teenager by a neighborhood watch member, and the bright light shone on the problematic way that dominant U.S. culture constructs black men—and even sixteen-year-old children—as monsters (Williams 2013).

3. For example, the lead author on Bawaka Country et al. 2015 and 2016 is the land/place in Australia where Bawaka people live.

4. Another key argument Wynter makes is that this Western bourgeois ethnoclass "genre of human" has shaped the development of knowledge in the Global North/West and has therefore also physically shaped neurocognitive pathways in the brains of people raised and/or educated in the West. These brain configurations continue to reproduce and contribute to continuing overrepresentation of this particular genre of human, even among activists wanting to create radical social change.

5. I explicitly state this because I know, from teaching race to seventy-five young people each year, and in other forms of outreach to the public, that the idea that there is genetic difference between "whites" and "blacks" as groups is commonly held in the United States today. For more on how there is no genetic basis for racial categories, see Marks (1995), Roberts (2011), and Templeton (2013).

6. The scholarship on race that builds on Omi and Winant is enormous and enormously varied, here I just list a few examples.

7. This line of thought could lead into the large bodies of scholarship on creolization (for two entries in that enormous literature, see Balutansky and Sourieau [1998] and Stewart [2007]), Afro-Creole cultural production (Burton 1997), and work that builds on Peter Wilson's articulation of the reputation–respectability dichotomy in his classic

ethnography of San Andres (Wilson 1973); and is echoed in scholarship on Caribbean linguistic production (Abrahams 1983)

8. The word "yam" is sometimes applied to the sweet potato in the United States, but the yam in the Caribbean and West Africa are larger, firmer, and white-fleshed and belong to a different plant family.

9. An interesting development in this research is represented by projects that explore how the agentive capacity of the more than human engages in worlding, contributing to the generation of human being in the process (e.g., Hathaway forthcoming; Tsing 2015).

10. In this endeavor, I have been influenced enormously by bell hooks in her discussion of radical openness on the margins, and learning from her in dialogues she has held at Southwestern University (SU), and in continuing conversations with my colleagues here at SU, especially Eric Selbin.

11. This is certainly not unique to rural Belize. My native Hawaiian nephew-in-law, Bernard Benanua, who comes from a small fishing community on one of the smaller Hawaiian islands, shared with me and his partner, my Irish American niece, Brighid Doherty, a similar sensibility. Both my niece and I perform a particular college-educated professional middle-class white positionality by asking questions (we see it as "being polite by seeming interested"), a practice that my nephew and his fellow rural Hawaiians, and my rural Creole husband, find off-putting and nosy.

12. This worldview also pervades academia much more powerfully than many think.

13. William Salmon studies Belizean Kriol (Salmon 2014, 2015; Salmon and Menjivar 2014, 2016), and Ken Decker has described the grammar of Kriol (Decker 2005). See also Geneviève Escure (2013).

CHAPTER 2 — HEWERS OF WOOD

1. "Ezekiel" is a pseudoymn. Almost all of the names in this book are pseudonyms. My husband, Elrick, is an exception.

2. Many records were lost when Hurricane Hattie hit Belize in 1961.

3. I will not cite sources for every historical detail that is common knowledge in and of Belize. My knowledge of Belizean history has been gleaned from my readings of Ashcraft (1973), Bolland (1977, 1988), Bolland and Shoman (1977), Campbell (2011), Grant (1976), Leslie (1995), and Shoman (1994). I have also looked at Dobson (1973) and Naylor (1989). For the very earliest days, see Burdon (1931), Campbell (2011), Dampier (1927 [1699]), Quijano (1944), and Uring (1726).

4. These seasons have been less predictable in recent years—sometimes there is no longer a dry season in August, sometimes the April dry season now stretches into June. Rural Belizeans point to climate change to explain these shifts.

5. The word "barcaderes" is undoubtedly an Anglicization of the Spanish "embarcadero," or landing spot on an inland waterway.

6. Buccaneer originates from French *boucanier* woodsman, pirate (in the seventeenth century West Indies), from *boucaner* to smoke meat, from *boucan* wooden frame for smoking meat, from Tupi words meaning "to be roasted." Its first known use: 1686 (www.merriam-webster.com).

7. Bulmer-Thomas and Bulmer-Thomas (2012) note Uring's cantankerous character, and suggest that Uring's commentary be read with that in mind. Dampier was hired in Campeche, today part of Mexico and a place where British buccaneer history is less apparent.

8. Exact numbers here are unclear, the censuses surely missed many people, enslaved and free, as Nigel Bolland (1977) notes.

9. In the very earliest descriptions of the region there are occasional mentions of apparently free people of African descent (these could either have been people who escaped slavery or individuals from the African continent who hired themselves on as sailors). For example, in 1676 Dampier refers to a "Negro" whom he hired to cure him of "guinea worm," and whom he paid with a "white cock" (Dampier 1927, 188). Dampier refers to enslaved blacks as "Negro-slaves" throughout the rest of this account.

10. St. George's Caye is also famed for the mythic battle between England and Spain that in the late eighteenth century resulted in a solid British claim to Belize. The role of this battle in the development of Belizean Creole culture and a sense of Belizean nationalism is fascinating and complicated (Judd 1990, 1992; Stone 1994). Suffice it to say that its Anglophilia serves to mask and downplay antiblack racism, the violent history of slavery in Belize, and it waters down the legitimate claims of Maya and Garifuna populations to full economic and political enfranchisement in Belize, and nurtures sentiment against Central American refugees.

11. The dye that was used in the burgeoning textile industry in Victorian England is today used as a stain in biological and medical applications, and also still in some cloth dying (see http://www.abbeycolor.com/logwood-extract.php).

12. Mavis Campbell makes a compelling argument that the only reason Britain could defend any Central American lands from Spain was because of its alliance with the powerful Miskito, who had successfully kept Spanish forces from conquering them over the years (Campbell 2011).

13. Enumeration of populations along the Central American coast at this time fluctuate widely and are of questionable accuracy. Furthermore, residents of Belize routinely left Belize for the safety of the British territory on the Mosquito Shore every time that Spain raided woodcutting camps in Belize; knowing who lived primarily where is difficult.

14. On reading Belize's history through maps, see Hoffman 2014

15. This sense of connection is also accompanied by a fear of "waika." In Belize Kriol, waika refers to the mixed indigenous and African populations of the Mosquito Shore who know "bad/black magic," are very knowledgeable about the "bush," and are very dangerous. Rural Belizeans tease each other about having waika blood, for instance.

16. What was happening here in these marginal parts of Belize is not dissimilar to what Stark (2015) describes for Puerto Rico in his analysis of the "hato" economy, a combination of livestock ranching, foodstuff cultivation, and timber harvesting that was commonplace and widespread in the eighteenth century, in which the institution of slavery was more flexible than in plantation contexts.

17. Marooned slaves, and survivors and shipwrecked slave ships fled to the Mosquito Shore. Different indigenous groups living in the area welcomed the formerly enslaved, and offered them a safe place to live. Intermarriage was common and a shared Miskito culture emerged out of these populations over the years. The Miskito became known for their independence and political and military savvy.

18. Britain abolished the slave trade in 1807, abolished slavery in 1834, but allowed slave owners to retain their slaves as "apprentices" depending on their age for the next four to six years. All apprenticeships ended in 1840.

19. Backlanding is a significant point in Belize, as it marks the point where the Old River system comes closest to meeting the New River, and has been on a major track that has long existed in Belize (see Figures 2.1 and 2.2).

20. The "th" sound in English is typically a plain "t" in Belizean Kriol.

21. District lines were later redrawn. Settlements that had been contained within Orange Walk District became part of Belize District. Today, the entire area I study is encompassed within the Belize District.

22. Elijah Bonner was my husband's great-grandfather, my children's great-great-grandfather.

23. The Garifuna came into being when a mutinied ship of enslaved Africans landed on St. Vincent and intermarried with the Carib Indians who were living there at the time, hence their other moniker: Black Carib. The Garifuna language is rooted in West African languages, Carib, Spanish, and French.

24. This is in flux however, as the Garifuna have more seemingly "authentic" cultural traditions, and have produced the music that has most put Belize on the cultural map (at least that of world music): punta rock.

25. The distinction between Maya and Mestizo in the late 1800s in Belize is a complex and interesting story (see Graham, Pendergast, and Jones 1989; Jones 1989; or, more generally, Bolland 1977, 1988; Shoman 1994).

26. Colorism is not limited to within Belizean Creole populations—similar processes obtain among the Maya and "Spanish," or Mestizo, populations in Belize. A Mayan last name precluded landownership, Spanish last names among Yucatec Maya (who only in very recent years have reclaimed their once-rejected Mayan identity) are associated with landholding; and light-colored skin, hair, and eyes even more so (see Judd 1992; Stavrakis 1979).

CHAPTER 3 — BUSH

1. As much as I would like to posit an opposite Kriol term to "backwaad," there is no such term, rural Belizeans will use the word "backwaad" often but there is not one particular word routinely used as its opposite. The most likely candidates include: civilized, modern, "like States" (like the United States).

2. The meaning of "the bush" shares similarity to the "country" of Raymond Williams's groundbreaking treatise on the country and city (Williams 1973) and follows what Aisha Khan (1997) rightly claimed has remained under-analyzed in the Caribbean (and multicultural postcolonial new world societies, generally): how identities are formed in relationship to the rural–urban dyad.

3. The word originated in Britain in the 1300s, and is very similar to an Old High German word for forest (busc) (https://www.merriam-webster.com/dictionary). Interestingly, in South Asia, the term "bush" is not as prevalent. Instead "jangal"—a Hindu word—served that purpose (Dove 1992). A thorough historical analysis of the term "bush" and how it has been employed throughout the British Empire has yet to be written.

4. Rivke Jaffe, in a discussion of urban environmentalism in the Caribbean has surveyed Jamaicans and Curaçaoans on their ideas about wild lands/wilderness and describes urban Caribbean attitudes toward "bush" (and its Dutch cognate, "mondi"). Although Jaffe is not aiming to unpack the meaning of "bush," the attitudes she finds are similar to what I have found in Belize, the bush/mondi "is seen simultaneously as

dirty and dangerous, a place of beauty and peace, a national symbol and an economic resource" (Jaffe 2008, 224). She also notes that urban Jamaicans dislike bush more than people in Curaçao, and she attributes that to the relative rurality of Curaçao. In both countries, however, cleaned-down development is better than bush.

5. I have never heard the word used to describe anyone who is not Creole. Although Belize Kriol is the lingua franca of Belize, the term "bushy" to describe a person appears to be limited to Creole (black, and not Garifuna) Belizeans. This is yet another dimension of the work the word does in race talk.

6. The idea of "haadness" is very similar to ideas of being tough in Jamaica (Ulysse 2007), and people in Cameroon see white people as soft in similar ways (Nyamnjoh and Page 2002). The term "bakra," which means white in Belize and elsewhere in the Caribbean (and even the southeastern United States) likely has an origin in Ibo, as a term meaning white man and/or "he who surrounds or governs" (Cassidy and LePage 2002, 16).

7. This, on the other hand, goes along with racial discourse that casts blackness as more animal-like, more physical (and therefore able to handle physical stress), but turns that on its head in terms of value. These associations showed up in a 2016 survey of medical students in Virginia in which roughly half of the students held the false idea that black people experience less pain than people of other races (Hoffman et al. 2016).

8. This set of meanings overlaps with the idea of competence I describe in chapter 4.

9. See Khan (1997) for a similar discussion of rurality in Trinidad.

10. The taste of these home-processed seeds is markedly different from the industrial cashew seeds that companies like Planters sell, and most Belizeans much prefer the stronger flavored Belizean cashew seed to what they consider the wan and flavorless industrial variety.

11. Wilk tells the story of this change—and of the role of colonialism, class, and eventually an emerging nationalism in shaping the meaning of gibnut (and other foods) in Belize (Wilk 1999, 2006).

12. The phrase "village life" itself is popular and used often. It was also the title of a Belizean television series that created video features of different villages throughout the country. Relatedly, a popular radio program called *Rural Talk* celebrated rural Creole Belize living during the mid-2000s (on which the term "new neega" was not used infrequently, and generated many chuckles among all listeners and the talk-show host).

13. Belize Great Belize Productions Channel 5 always creates direct transcripts of its interviews—the partial Kriol/partial English of this interview is authentic and is how rural Belizeans communicated with me during formal interviews. When we were just chatting, rural Belizeans more commonly spoke mostly in Kriol.

14. This is, of course, part of the whole way that aesthetics, beauty, and skin color convene. A growing literature on colorism explores these linkages (see Norwood 2014).

15. This is expressed not only at the level of cultural meanings but also in Belize's land laws. One can claim ownership and receive a deed title, or leasehold right, to land if one has "improved" it—that is, cleared bush.

16. The word "picado" in Spanish means choppy, chopped up, broken—these pathways are so narrow and small, they themselves are choppy. Picado is commonly used by Creole people to describe small narrow pathways in bushy areas.

17. This may well be common in rural areas (especially in the Americas): small-scale fishermen in Puerto Rico (Griffith and Valdés Pizzini 2002) and lobster fishermen in Maine (Acheson 1988) are also known for sharing this sentiment.

18. There has been a recent florescence of scholarship on ideas of freedom in the Caribbean (see, e.g., Bonilla 2015; Roberts 2014, 2015).

19. In Belize, this has often meant simply living on a plot of land and passing it down through the centuries, not necessarily establishing legal title to it.

20. This does not mean that they do not also sometimes seek out government attention and assistance—improved roads, provision of electricity, and most recently, water service. One of the most heralded ancestors in Crooked Tree was a village leader who was able to very effectively negotiate with the Belizean government to ensure these kinds of provisions (and an exception for Crooked Tree to national deer-hunting laws).

21. One interesting twist on this sentiment is that village life is seen to be safer from crime than life in Belize City or in the urban United States where most Belizeans are located. This sense grew between 1990 and 2017, but now even some villages are also sites of murder and robbery and are not as safe as they were. The drug and small arms trade across the Americas and the gang and cartel political, economic, and social organizations that have emerged as a result of this trade, along with the lack of other economic opportunities, have created this shift.

22. In *Purity and Danger*, Mary Douglas makes the widely accepted point that dirt, or anything tabooed, is "matter out of place" (Douglas 1988). For rural Belizeans, I contend that the most critical "matter out of place" in the yard is unkempt nature—or bushy-ness—removing bushy-ness is central to being a rural Belizean.

23. A few individuals in the village opt not to chop their yards as much, preferring to live more in harmony with nature and to build up wildlife populations. This appears to have two different impetuses: involvement in the global circuit of nature tourism (one villager is a guide who travels broadly offering guiding services); and the florescence of interest in rural Belizean Creole practices as the root of Belizean culture.

24. The speaker may have used the word "jungle" either to "speak," which in Kriol means to speak English, rather than Belize Kriol, which people tend to do at village council meetings, or he may have used the word to emphasize even more how terrible it looked. Jungle is *not* a commonly used word in Belize.

25. Belizeans are not dissimilar to people in the United States in this way (Robbins 2007).

26. Various scholars have studied the Caribbean yard, but the most pertinent analysis for my argument is made by feminist theorists who identify the yard as a place where women act as creative agents, generating Afro-Caribbean culture through creating their yards (Besson 1987, 1993; Besson and Momsen 1987; Burton 1997; Momsen 1993). They assert this as a critique of other scholars who suggest that Afro-Caribbean culture—and particularly the more "Afro" than "Euro" dimensions of that culture—is most nurtured in the street (Brana-Shute 1979; Wilson 1973). These arguments are part of the debate about "reputation" versus "respectability" begun by Peter J. Wilson (1973). "Reputation" is seen as an Afro-Caribbean value of resistance to European norms and "respectability" as an accommodation to European norms. The dichotomy correlated to the spatial antinomy of the street on the one hand, and the house and yard on the other (Wilson 1973; see also Abrahams 1983; Brana-Shute 1979; Douglass 1992; Mintz 2010).

27. Historically and still today, land has been kept chopped low with machetes, which are standard tools found in all rural Creole households. More recently, gas lawn mowers have become popular to help keep grass cut short, but machetes are still used on woodier brush that grows quickly in the subtropics and cannot be cut with a conventional lawn mower.

CHAPTER 4 — LIVING IN A POWERFUL WORLD

1. If they had been in a traditional wooden rural Creole dory, it would not have broken, and even if overturned, it would have remained floating nearby the men.

2. An early example of this kind of ethnography comes from Stephanie Kane, whose account of development in Panama flows from the thinking and worlding of the people she studied (Kane 1994).

3. This is also partly about the languages in which knowledge seen as legitimate is produced. Carolyn Cooper offers an interesting intervention here. She published a literary criticism article in Patois, the Creole language spoken in Jamaica (Cooper 2015). When I assigned this in an undergraduate class on the Caribbean, it was difficult for the students to read and they understood my point. Similarly, Gina Ulysse's book *Why Haiti Needs New Narratives* (2015), which in one volume contains her essays in three languages: Haitian Kreyol, English, and French, powerfully intervenes in this problem of language and knowledge. I also assigned this, and both I and my students came away with a deeper understanding of the coloniality of academia, of the Euroversity.

4. "Dory," a Kriol word for canoe, is ideally a dugout carved from the trunk of a tubroos or mahogany tree. Knowing how to make a dory is important Creole knowledge, too. For the past decade or so, there has been a big dory race down the Old Belize River that attracts international competitors and an international crowd and celebrates Creole dory knowledge all at the same time.

5. Belizeans most prefer making tea with Brooke Bond loose black tea leaves in a pot of boiling water. They then mix in sweetened condensed milk and sugar.

6. Programme for Belize, one of the most prominent environmental organizations in the country, operates a very large protected area that stretches toward Guatemala from these villages (see chapter 6).

7. The capacity for animals to radically transform ecosystems is just beginning to be acknowledged in western/northern scientific studies of environmental change (e.g., wolves changing rivers in Yellowstone); ironically, this recognition comes during the Anthropocene. *How Wolves Change Rivers* (https://www.youtube.com/watch?v=ysa5OBhXz-Q).

8. Alan Rabinowitz concurs with this, and broadens this assertion to anywhere in the jaguar's range, having found no documented evidence of jaguars attacking humans without provocation (Rabinowitz 2000).

9. Man-eating tigers in India in British colonial history, and both the African and British experience of lions and leopards in Africa (all three of these kinds of cats are more likely to attack humans than are jaguars) have certainly contributed to the highly developed fear of jaguars in Creole culture.

10. He was not only worried about jaguars hurting me—though certainly that was our conversation at the time—but also about Tata Duende and other spirits (as I discuss below). I learned about all his concerns in the summer of 2014—ten years later. Our brother-in-law went missing in Belize—in a particularly spirit-full part of the Belizean

bush. As we tried to understand the disappearance, my husband said that Tata was one of the reasons he was so worried I had walked out alone through that ridge. In particular, where I had felt funny, is ridge (or a forested area) known to be home to many bad spirits.

11. Though the amount a hunter could get for a skin has never been very large—especially once it became illegal to kill jaguars and trade their skins internationally (interestingly, illegality did not spike the value of the skins).

12. Tata Duende is not at all unique to rural Creole Belize and appears to have some kind of syncretic origin between indigenous American (Mayan and other groups) and southern European beliefs (and there are similar creatures as far as the Philippines—suggesting Spain as an important source). Tata Duende is common throughout Central America, though his exact name, precisely how he looks, and what he is specifically capable of doing varies from place to place, language to language, culture to culture. His name in Spanish translates simply to "grandfather owner." Many of the written descriptions of him note that he is the protector of the forests—of the animals and plants that live in the forests. This dimension of him was not emphasized to me by Belizeans I know. In my understanding of him, he is not quite that benign.

CHAPTER 5 — ENTANGLING THE MORE THAN HUMAN

1. J. K. Gibson-Graham is the pen name of two feminist political economists who even in their name enact an alternative version of academia.

2. Depending on how these phrases are uttered, they might refer more to balancing sociality and earning cash, negotiating larger political and economic variables in the struggle to survive, or cobbling together small jobs to earn "just enough" cash to live. Belizeans also say, "I de rite ya de fite life" (I am right here fighting life/surviving)—which more directly connotes negotiating the difficulties posed by economic, political, and cultural structures and institutions. See also Browne (2004).

3. How nativeness, or indigeneity, of fish, and what it means to be Creole work together is of course at play here and will be addressed in more detail in chapter 7.

4. Writing about turtles is challenging to me for a number of reasons. First, part of me is a deep nature lover, a conservationist, I am anti-speciesism—the hicatee has every right to persist—I am not sure how endangered it is. My husband and many Belizeans are convinced there are no fewer hicatee than there used to be. Nevertheless they sometimes also express concern for the turtle's well-being, and worry that the turtles are overharvested. They might be right, and I typically trust the assessment and knowledge of rural Creole people over the conservation biologists who come to Belize often but only for short visits. Still, I read the biologists' assessments and I know of many rural Creole people who hunt these turtles. As with any endangered species, their value increases as they become more endangered (see chapter 6). Selling a turtle is becoming a great way to make a lot of cash. I might hunt and sell them too if I lived here most of the time, but I would not likely eat them. I am not a particular fan of hicatee meat—I can eat it, but I do not love it the way I do the common-as-dirt hamadilly, for instance. The other complicated part of thinking about turtles for me is that turtles provided me with one of first challenges of living differently with the more than human in Belize. A turtle had been stored in the house where I was staying for several weeks, placed on its back, alive in standard storage mode. One day, it was brought out to be butchered. Its head was cut off and its shell removed, but the body kept moving. This, of course, was completely

normal—that is how bodies die, but I had never witnessed an animal being butchered for human consumption before, and the moment of watching that was painfully seared into my memory.

5. Historically, these nets were homemade, but fewer and fewer people are masters of this art, and indeed it is technically illegal to hunt hicatee with nets today (see chapter 6).

6. Of course, not all men hunt, and a few (but not many) women occasionally hunt.

7. I know of no Creoles who have eaten, or would eat, a mountain cow.

8. See an interesting discussion of just this experience written by a rural black South Carolinian man (Lanham 2016).

9. I was so certain that he sold his catch, that I talked with my husband about why Ozzie was not telling me the truth about selling the game he hunted. My husband laughed and said that it was true, he does not sell his catch; he said he and his sisters get mad at Ozzie about that all the time. I had never noticed, because we are in his very close circle of family and friends and I expected him to share with us, but had no idea that he shared with lots of others too!

10. This is not dissimilar from the avid deer hunters in Texas who feed deer corn in their yards, but then go out to their leases to kill them on the weekends in the fall.

11. I find this issue very interesting. I wonder about the similarities between all forms of intimacy—sexual, maternal/paternal, eating. I think about cats who purr and lick and bite gently, who sometimes gently carry their prey after they have killed it, before ripping it apart to eat. I think about how some male mammals bite females before copulating. The boundaries are slippery between types of animal mingling.

12. Although cattle that are not fenced routinely roam through the village as well, and this has been a source of discord in some villages (see Johnson 2015).

13. This is no easy feat from my perspective. I have co-owned cattle in Belize for seventeen years and spent months among them, but cannot tell them apart. To me they are either white or brown. That is all I am capable of distinguishing.

14. This is beyond the scope of the book, but worth noting: midwives, who until the 1990s delivered most of the village women's babies, have both deep and wide-ranging knowledge of many different plant-based healing remedies.

15. After Hurricane Irma flooded parts of Miami in September 2017, Haitian women who knew how to make fire hearths were critical providers of food to their neighbors during the days residents were living without electricity. It is an important bit of knowledge and skill in multiple contexts.

16. Although space precludes my developing this argument further, rural Creole economic activity also disrupts conventional reading of the human as first and foremost a laborer or producer. Indeed, this is a critical point that rural Creole people often make: "I de rite ya di do nuttin" (I'm right here doing nothing). This represents a deep proclivity to be one's own boss, to engage in a wide range of activities to maintain life. Ideally, they are far away from people telling them what to do, away from centers of power. This being-ness reflects Caribbean-wide sensibilities, and also disrupts European/U.S. white male *homo oeconomicus* normative subjectivity (see Wynter 2003)

17. This mix of Creole and English was common in my more formal interviews with people.

18. Another related body of scholarship notes that what people do (what labor they engage in) generates their identity. Scholars studying indigenous populations (Nelson

1986 [1973]) note this point, but so has White (1995) in his study of loggers in the Pacific Northwest. Astuti (1995) observes that ethnic identity in Madagascar changes when people shift their primary subsistence activity.

19. Describing Creole languages is politically complicated. Caribbean Creole languages emerged out of the (violent) encounter between European and West African languages. In Belize, the Kriol language has roots in the English and West African languages. Any Creole language contains within it a continuum—people speak versions that have either more African or more European content and/or structure. In addition, some Creole languages, like Bajan, have fewer recognizable African features than do other Creole languages (for example, Jamaican patois). Historically, the way in which people spoke Creole, where on the continuum the language they were speaking at any given moment could be located, was defined by how distant a speech pattern was from the European contributing language. This gives primacy to the European language. All people who speak Belizean Kriol use the language variably and include more African or more European elements depending on the context.

20. http://mrpetersboomandchime.bandcamp.com/track/crooked-tree-and-burrel-boom. The song also mentions Burrell Boom—a larger village located closer to Belize City, the urban heart of Belize (see Map 1.2).

21. https://www.youtube.com/watch?v=slaFmRkniVg. The Bonner/Banner family traces its origins to two brothers—the first settled in Lemonal Village in Belize District at some point in the eighteenth or nineteenth century, the other settled in Camalote Village in Cayo District. Family members would have traveled between the two communities on the Belize Old River.

CHAPTER 6 — WILDLIFE CONSERVATION, NATURE TOURISM AND CREOLE BECOMINGS

1. I was not there to ask him what he meant by the word "livity," a term I find compelling. He may simply have meant livelihood, which it has come to mean in urban slang. However, the term is rooted in Rastafarianism and suggests a broader meaning that encompasses spiritual and natural well-being. For more on this, see chapter 8.

2. Colonial activities in Belize reflected the colonial conservation regime (or technologies for controlling the natural environment) that Britain was developing elsewhere in its empire (see Anderson and Grove [1987] on Africa, and Tucker [1982] on India).

3. The degree to which areas designated as protected are actually protected or merely "paper parks" is debated. The government of Belize has limited monies for conservation. Some protected areas are actually managed by NGOs as a result (this is true of the protected areas in this study) (see Johnson 2015).

4. Early maps clearly indicate that this particular area was the centerpiece of early settlement and resource extraction, but in the twentieth century, this place became less critical and central to the economy of British Honduras. Yet a certain sense of entitlement and privilege still obtains among residents of this part of Belize because of their ancestral history.

5. See Michael Hathaway (2013) for an analysis of conservation and human-environment relations in Southwest China. He intriguingly shows how rural Chinese as well as scientists, policymakers, and government officials, all sometimes intentionally reached out to attract global processes to themselves.

6. Given that the water levels are anything but constant here, this has always been a point of dispute.

7. Villagers work as wardens and sometimes in the Belize Audubon Society offices as well. In this way, Crooked Tree fits with some of what Nora Haenn (2016) addresses in her analysis of conservation in the Yucatan—conservation officials and workers are not necessarily elite or external, but villagers still construct Audubon and Programme for Belize as external and white.

8. Government is called in sometimes: if a physical fight results in a hospital visit, charges might be filed; if someone has stolen one too many cows, police may be called in; but very often people do things their own way here, taking advantage of being far "off-the-main."

9. This point thus echoes debates around "environmentality" (e.g., Agrawal 2005; Cepek 2011; Medina 2015). I prefer the frame of assemblages to make sense of what I see in Belize—the ways of thinking and being I document here are not so much an effect of techniques of rule, per se, but rather are sensibilities more freely embraced and discarded over time.

10. "Shame" is a central Kriol conceptual domain and emotional state of being that is shared throughout the Afro-Caribbean. It is a state of being that people go to great lengths to avoid. Some examples of what might bring on this state include: if a person is "cussed out" and cannot "quarrel" back, if someone is caught doing something they should not do, or, if someone who feels lacking in good table manners is forced to eat in front of people of higher status.

11. "Da sin" is a common Kriol phrase, similar to "It's a shame," or "It's a sin" in the United States. It does not particularly refer to religious beliefs.

12. Some children still pluck and cook their catch in "bush camps" they make on the edges of villages.

13. On June 30, 2016, my husband received a call informing him that one of his calves had been attacked by a jaguar, but the calf was still alive at that point. The jaguar track, right in the yard of his sister's house on the road into Lemonal, immediately next door to our house, was huge.

14. In 2005, the density rate in this part of Belize was 11 / 100 km^2 (Meerman 2005), compared to Amazon basin, 4.4 ± 0.7 / 100 km^2 (Tobler et al. 2013).

15. While I might question some individuals' claims on this front—how can you be sure it was a tigah that took the calves?—I trust Mr. Raymond—there are some consistent and easily identifiable ways that mark a tigah kill (Roberson 2015).

16. This mode of interacting was historically not the way that Audubon (or government officials) dealt with community members, and may have been as much a result of one particularly zealous warden's tenure as much as anything else.

17. The person who made the report would likely have been someone with more conservationist leanings or someone very tightly tied to the ecotourism industry.

18. Various initiatives are under way in Belize to try to find ways that rural Belizeans can continue raising cattle while healthy jaguar populations are maintained (Roberson 2015; Steinberg 2016).

19. This followed shortly after the popularity of fancy bird feathers in hats—heavy trade in things animal, animal parts, for personal attire/accessory as a Eurocentered way of relating to the more than human in the late nineteenth to mid-twentieth century (see West 2006, esp. ch. 1).

20. This, however, was not true, at least according to official figures. The 1991 census and I agree that the population of Crooked Tree was around six hundred. However, the debate over the census number raises the whole complicated issue of who lives in Crooked Tree. If the census count included individuals who only live in the village for a few months each year, then the number is much higher. And the ties between people residing in Crooked Tree and those residing elsewhere further complicate questions of boundaries (see Wilk and Miller 1997).

21. ERI itself is Belizean, but some faculty as well as funding come from the United States (exemplifying an interesting melding of middle-class urban/Belmopan Belizean conservation and U.S. funding).

22. These were a university student, a government official, and an NGO representative—a typical mixture for conservation representatives—and all relatively elite compared to the villagers, although a few villagers have children who could easily have been one of these representatives. Belize has a relatively large number of conservation-related jobs in government and NGOs, and these are high-status jobs.

23. In this meeting, the village leader spoke in English, while the presenters often used Kriol. Villagers work to claim status through speaking English. The university students tried to create a connection to community members by speaking Kriol (see Johnson 2015, 85).

24. The biggest threats to the well-being of the more than human in this part of Belize is the possibility of large-scale conversion of the subtropical forests, scrubland, and savannahs to agricultural production—monocrop planting, large-scale cattle ranching, and the like. Hunting pressure and small-scale farming is not as problematic, but see Foster et al. (2016).

25. Numerous feature articles in newspapers from around the United States serve as a testament to this. They provide exciting and romanticized accounts of mostly men (but an occasional woman) setting out on explorations of the jungle, on jaguar hunts, or hunting for Maya ruins.

26. A white Englishman living in El Salvador, Sidney Stadler, and Sir Ronald Garvey, a white colonial officer in British Honduras, met and discussed the issue of "beautifying the highways" of Belize in the late 1930s. In a letter to Garvey, the British El Salvadoran explained how he formed a *comité de ornato* with Salvadorans to beautify their roadsides and city parks with native flowering trees. In 1951, and following upon these earlier conversations, a committee was established in Belize to "consider how best certain areas in the Colony should be preserved as special areas of beauty or interest." The perspective informing this endeavor is white, elite, and colonial British, and coincides with the beginnings of the development of mass tourism. This committee called itself the "Committee for Preservation of Areas of Interest and Beauty" (CPAIB) and was initially made up of three colonial officers: the director of surveys, the conservator of forests, and the archaeological commissioner. These three committee members visited a wide array of potential sites "of interest or beauty" throughout the colony and suggested ways to make them more attractive (Minute Paper 632 of 1951).

27. Curtis J. Prock, the white U.S.-based tiger camp owner, was famed for his hunting skills but was also charged with running illegal canned trophy hunts in the United States. Some of the jaguar sightings/kills in the 1960s and 1970s in Arizona are thought to be jaguars he captured and then released for hunters there (Brown and Thompson 2010).

28. The lodge was run by Barothy for many years, and then by Keller, but called Wade Bank by locals—reflecting its longer history of being home to the Wade family, whose descendants still populate the area.

29. "Baboon" is the Kriol word for black howler monkey.

30. The two most successful of these were relatively light-skinned, which likely is not coincidental.

CHAPTER 7 — TRANSNATIONAL BECOMINGS

1. The wait to adjust status can take up to twenty years, and costs a large amount of money.

2. This place on the edge of the Caribbean Sea has been marked by human modified movement and exchange for millennia.

3. The movement of rural Belizean nature-goods is not limited to food, and can enroll non-Belizeans as well. For example, "Bush medicine" to improve female fertility came to Texas with Elrick to help a friend reproduce.

4. Barbecue has its origins in the foodways of indigenous peoples of the Caribbean islands and was swiftly adopted by the British shortly after they settled in this area, including those in what is today Belize.

5. Fort Hood, in Killeen, Texas, is home to many Belizeans, who like other immigrants from the Caribbean are disproportionately represented in the U.S. military.

6. Rice and beans is *the* national Sunday dish for Belizeans, and is an important dish for many Latin American and Caribbean peoples, each nation and cultural group preparing it their own way (Wilk and Barbosa 2012). The fried plantain itself is worthy of note. This was not everyday plantain from our local supermarket, but had come from an Asian supermarket thirty miles away in Killeen, next to Fort Hood, with its large population of immigrants. Elrick had traveled there to buy a bucket of salt-cured pig tail that he puts in stewed red kidney beans (he sent some pig tail home with our visitor), and saw the plantain. Belizeans work hard to find the foods that taste most like home.

7. I am most active on Facebook and encounter a wide range of Belizeans (age, location, gender) on this platform. My sense is that Facebook is the most commonly used platform across the diaspora, but I am not certain of this.

8. The linguistics of Facebook posts are really interesting. When people choose to write in Belizean Kriol, for which a standardized spelling system has only recently been created, how they choose to use it, and spell it, are all worthy of study. I myself sometimes post using Kriol, sometimes not. I am not sure exactly when and why I choose which language to comment in.

9. That they call it Coolie Ranch is interesting and funny in and of itself. They are indeed East Indian Trinidadians, the derogatory term for East Indians throughout the Caribbean is "Coolies," although many take that moniker with pride as well.

10. Although being bushy, or having rural roots, does not automatically confer hunting skill, most people who are from the bush are more skilled at hunting than city folk.

11. Nicknames are an interesting and important linguistic phenomenon in Belizean Kriol. Many people have nicknames, and are only, or mostly, known by them. Indeed, Elrick is known much more by his nickname "Seedy," which was supposedly jokingly given to him by his mother when he was a toddler. His mother had very long straight "good" hair, and Elrick had very kinky hair, like "cold seed," Kriol for goosebumps, on his head, and so he was called Seedy. A community leader in Crooked Tree was trying

to put together papers that organized cricket teams; the team entered from Lemonal was led by "Elrick Bonner," the leader said she had no idea that this was "Seedy" until she asked several people who this "Elrick" was.

12. Lionfish are a Caribbean-wide problem (see Moore 2012).

13. The question of what constitutes an invasive, or "alien," species is complicated (see Fortwangler 2009; Knudsen 2014; Pearce 2015; Richardson 2011). The idea of a "species," a defined and bounded type of being, that is "native" to a "place" is also, as recent scholarship has been making clear, fraught with underlying ideas of racial purity and fixity (Kim 2015; Subramanian 2001).

14. Western/Euro/Anglo science knows tilapia to be omnivorous and capable of eating baby fish, but not exclusively or as voraciously as imagined. Rural Belizean fishermen know tilapia to be vegetarian. The tilapia in rural Northern Belize are likely vegetarian.

15. Interestingly, Ava is identified in this news story as a cook, but she has been a village council woman, has been involved in politics and a ladies group. She has had multiple roles and is a community leader.

Bibliography

Abrahams, Roger D. 1983. *The Man-of-Words in the West Indies: Performance and the Emergence of Creole Culture*. Baltimore: Johns Hopkins University Press.

Abranches, Maria. 2014. "Remitting Wealth, Reciprocating Health? The "Travel" of the Land from Guinea-Bissau to Portugal." *American Ethnologist* 41 (2): 261–275.

Acheson, James M. 1988. *The Lobster Gangs of Maine*. Hanover, N.H.: University Press of New England.

Adichie, Chimamanda Ngozi. 2013. *Americanah: A Novel*. New York: Anchor Books.

Agrawal, Arun. 2005. *Environmentality: Technologies of Government and the Making of Subjects*. Durham, N.C.: Duke University Press.

Allen, Paula Gunn. 1992 (1986). *The Sacred Hoop: The Feminine in American Indian Tradition*. Boston: Beacon Press.

Allewaert, Monique. 2013. *Ariel's Ecology: Plantations, Personhood, and Colonialism in the American Tropics*. Minneapolis: University of Minnesota Press.

Anderson, David, and Richard Grove. 1987. *Conservation in Africa: People, Policies, and Practice*. Cambridge: Cambridge University Press.

Anderson, Jennifer L. 2012. *Mahogany: The Costs of Luxury in Early America*. Cambridge, Mass.: Harvard University Press.

Anderson, Kay. 2003. "White Natures: Sydney's Royal Agricultural Show in Post-Humanist Perspective." *Transactions of the Institute of British Geographers* 28 (4): 422–441.

Ashcraft, Norman. 1973. *Colonialism and Underdevelopment: Processes of Political Economic Change in British Honduras*. New York: Teachers College Press.

Astuti, Rita. 1995. "'The Vezo Are Not a Kind of People': Identity, Difference, and 'Ethnicity' among a Fishing People of Western Madagascar." *American Ethnologist* 22 (3): 464–482.

Austin, Diane J. 1984. *Urban Life in Kingston, Jamaica: The Culture and Class Ideology of Two Neighborhoods*. New York: Gordon and Breach.

Baines, Kristina. 2016. *Embodying Ecological Heritage in a Maya Community: Health, Happiness, and Identity*. Lanham, Md.: Lexington Books.

Balutansky, Kathleen M., and Marie-Agnès Sourieau. 1998. *Caribbean Creolization Reflections on the Cultural Dynamics of Language, Literature, and Identity.* Gainesville: University Press of Florida.

Bandy, Joe. 1996. "Managing the Other of Nature: Sustainability, Spectacle, and Global Regimes in Capital in Ecotourism." *Public Culture* 8 (3): 539–567.

Barrow, Dean. 2008. belizemediacenter.org. Retrieved May 26, 2008.

Bawaka Country, Sarah Wright, Sandie Suchet-Pearson, Kate Lloyd, Laklak Burarrwanga, Ritjilili Ganambarr, Merrkiyawuy Ganambarr-Stubbs, Banbapuy Ganambarr, and Djawundil Maymuru. 2015. "Working with and Learning from Country: Decentring Human Authority." *Cultural Geographies* 22 (2): 269–283.

Bawaka Country, Sarah Wright, Sandi Suchet-Pearson, K. Lloyd, L. Burarrwanga, R. Ganambarr, M. Ganambarr-Stubbs, B. Ganambarr, D. Maymuru, and J. Sweeney. 2016. "Co-becoming Bawaka: Towards a Relational Understanding of Place/Space." *Progress in Human Geography* 40 (4): 455–475.

Beach, Tim, Sheryl Luzzadder-Beach, Duncan Cook, Nicholas Dunning, Douglas J. Kennett, Samantha Krause, Richard Terry, Debora Trein, and Fred Valdez. 2015. "Ancient Maya Impacts on the Earth's Surface: An Early Anthropocene Analog?" *Quaternary Science Reviews* 124: 1–30.

Beletsky, Les. 1999. *The Ecotraveller's Wildlife Guide: Belize and Northern Guatemala.* San Diego, Calif.: Academic Press Natural World.

Beliso-de Jesús, Aisha. 2015. *Electric Santería: Racial and Sexual Assemblages of Transnational Religion*. New York: New York University Press.

Bennett, Jane. 2010. *Vibrant Matter: A Political Ecology of Things.* Durham, N.C.: Duke University Press Books.

Benya, Edward. 1977. "The Art of the Carbonero." *Belizean Studies* 5 (6): 20–27.

Benya, Edward. 1979. "Forestry in Belize, Part I: Beginnings of Modern Forestry and Agriculture, 1921 to 1954." *Belizean Studies* 7 (1): 16–28

Berry, Kate A., and Martha L. Henderson. 2002. *Geographical Identities of Ethnic America: Race, Space, and Place.* Reno: University of Nevada Press.

Bessire, Lucas. 2014. *Behold the Black Caiman: A Chronicle of Ayoreo Life.* Chicago: University of Chicago Press.

Besson, Jean. 1987. "A Paradox in Caribbean Attitudes to Land." In *Land and Development in the Caribbean*, edited by Jean Besson and Janet Momsen, 13–45. London: Macmillan Caribbean.

Besson, Jean. 1993. "Reputation and Respectability Reconsidered: A new perspective on Afro-Caribbean peasant women." In *Women and Change in the Caribbean: A Pan-Caribbean Perspective*, edited by Janet H. Momsen, 15–37. Kingston, Jamaica: Ian Randle.

Besson, Jean, and Janet Momsen, eds. 1987. *Land and Development in the Caribbean.* London: Macmillan Caribbean.

Birth Records of Orange Walk District. 1913. Births from 1888 through 1913. National Archives of Belize, Belmopan.

Blaser, Mario. 2010. *Storytelling Globalization from the Chaco and Beyond.* Durham, N.C.: Duke University Press.

Blaser, Mario. 2014. "Ontology and Indigeneity: On the Political Ontology of Heterogeneous Assemblages." *Cultural Geographies* 21 (1): 49–58.

Blaser, Mario. 2016. "Is Another Cosmopolitics Possible." *Cultural Anthropology* 31 (4): 545–570.

Boardman, Robert. 1991. *International Organization and the Conservation of Nature*. London: Macmillan.

Bolland, O. Nigel. 1977. *The Formation of a Colonial Society: Belize, from Conquest to Crown Colony*. Baltimore: Johns Hopkins University Press.

Bolland, O. Nigel. 1988. *Colonialism and Resistance in Belize: Essays in Historical Sociology*. Benque Viejo del Carmen, Belize: Cubola Productions.

Bolland, O. Nigel, and Assad Shoman. 1977. *Land in Belize: 1765–1871*. Kingston, Jamaica: Institute of Social and Economic Research.

Bolles, A Lynn. 1996. *Sister Jamaica: A Study of Women, Work and Households in Kingston*. Lanham, Md.: University Press of America.

Bonilla, Yarimar. 2015. *Non-Sovereign Futures: French Caribbean Politics in the Wake of Disenchantment*. Chicago: University of Chicago Press.

Bonilla-Silva, Eduardo. 2013. *Racism without Racists: Color-blind Racism and the Persistence of Racial Inequality in America*. Lanham, Md.: Rowman and Littlefield.

Bonnett, Alastair. 1999. *White Identities: An Historical and International Introduction*. London: Routledge.

Boyce Davies, Carole. 2015. "Chapter 9: From Masquerade to *Maskarade*: Caribbean Cultural Resistance and the Rehumanizing Project." In *Sylvia Wynter: On Being Human as Praxis*, edited by Katherine McKittrick, 203–225. Durham, N.C.: Duke University Press.

Brahinsky, Rachel, Jade Sasser, and Laura-Anne Minkoff-Zern. 2014. "Race, Space, and Nature: An Introduction and Critique." *Antipode* 46 (5): 1135–1152.

Brana-Shute, Gary. 1979. *On the Corner: Male Social Life in a Paramaribo Creole Neighborhood*. Assen, Netherlands: Van Gorcum.

Braun, Bruce. 2002. *The Intemperate Rainforest: Nature, Culture, and Power on Canada's West Coast*. Minneapolis: University of Minnesota Press.

Braun, Bruce. 2003 "'On the Raggedy Edge of Risk': Articulations of Race and Nature after Biology." In *Race, Nature, and the Politics of Difference*, edited by Donald S. Moore, Jake Kosek, and Anand Pandian, 175–203. Durham, N.C.: Duke University Press.

Braun, Bruce. 2006. "Environmental Issues: Global Natures in the Space of Assemblage." *Progress in Human Geography* 30 (5): 644–654.

Braun, Bruce, and Noel Castree, eds. 2005. *Remaking Reality: Nature at the Millennium*. London: Routledge.

Brenick, Alaina, and Rainer K. Silbereisen. 2015. "Leaving (for) Home: Understanding Return Migration from the Diaspora." *European Psychologist* 17 (2): 85–92.

Bridgewater, Samuel. 2012. *A Natural History of Belize: Inside the Maya Forest*. Austin: University of Texas Press.

British Honduras Census. 1832. Belmopan: National Archives of Belize.

British Honduras Census. 1835. Belmopan: National Archives of Belize.

Brockington, Dan. 2002. *Fortress Conservation: The Preservation of the Mkomazi Game Reserve, Tanzania*. Bloomington: Indiana University Press.

Brockington, Dan, Rosaleen Duffy, and Jim Igoe. 2008. *Nature Unbound: Conservation, Capitalism and the Future of Protected Areas*. London: Earthscan.

Brown, David E., and Ron Thompson. 2010. "Lions, Tigers and Bears, Oh My! The Legacy of Curtis J. Prock." In *Arizona Wildlife Trophies*, edited by Richard L. Glinski, Duane J. Aubuchon and Bill Keebler. Mesa: Arizona Wildlife Federaion.

Browne, Katherine E. 2004. *Creole Economics: Caribbean Cunning under the French Flag*. Austin: University of Texas Press.

Bulmer-Thomas, Barbara, and Victor Bulmer Thomas. 2012. *The Economic History of Belize: From the 17th Century to Post-Independence*. Benque Viejo del Carmen, Belize: Cubola Productions.

Burdon, Sir John Alder, ed. 1931. *Archives of British Honduras, Volumes I and II*. London: Sifton Praed & Co.

Burton, Richard D. 1997. *Afro-Creole: Power, Opposition and Play in the Caribbean*. Ithaca, N.Y.: Cornell University Press.

Büscher, Bram, and Veronica Davidov, eds. 2013 *The Ecotourism-Extraction Nexus: Political Economies and Rural Realities of (un)Comfortable Bedfellows*. New York: Routledge.

Bush, Barbara. 1988. *Slave Women in Caribbean Society, 1650–1838*. Bloomington: Indiana University Press.

Campbell, Mavis Christine. 2011. *Becoming Belize: A History of an Outpost of Empire Searching for Identity, 1528–1823*. Kingston: University of the West Indies Press.

Carney, Judith Ann. 2009. *Black Rice: The African Origins of Rice Cultivation in the Americas*. Cambridge, Mass.: Harvard University Press.

Carney, Judith Ann, and Richard Nicholas Rosomoff. 2011. *In the Shadow of Slavery: Africa's Botanical Legacy in the Atlantic World*. Berkeley: University of California Press.

Cassidy, Frederic Gomes and Robert Brock Le Page. 2002. *Dictionary of Jamaican English*, Second Edition. Kingston: University of West Indies Press.

Castree, Noel. 2013. *Making Sense of Nature*. London: Routledge.

Cepek, Michael L. 2011. "Foucault in the Forest: Questioning Environmentality in Amazonia." *American Ethnologist* 38 (3): 501–515.

Checker, Melissa. 2008. "Eco-Apartheid and Global Greenwaves: African Diasporic Environmental Justice Movements." *Souls* 10 (4): 390–408.

Chevannes, Barry. 2001. "Jamaican Diasporic Identity: The Metaphor of Yaad." In *Nation Dance: Religion, Identity and Cultural Difference in the Caribbean*, edited by Patrick Taylor, 129–137. Bloomington: Indiana University Press.

Clarke, Kamari Maxine, and Deborah A. Thomas. 2006. *Globalization and Race: Transformations in the Cultural Production of Blackness*. Durham, N.C.: Duke University Press.

Clary, John Vincent. 2014. "Digital Geographies of Transnational Spaces: A Mixed-method Study of Mexico-US Migration." Masters thesis, University of Texas at Austin, Department of Geography.

Cleghorn, Robert. 1939. *A Short History of the Baptist Missionary Work in British Honduras: 1822–1939*. London: Kingsgate Press.

Cohen, Jeffrey H., and İbrahim Sirkeci. 2011. *Cultures of Migration: The Global Nature of Contemporary Mobility*. Austin: University of Texas Press.

Comitas, Lambros. 1973. "Occupational Multiplicity in Rural Jamaica." In *Work and Family Life: West Indian Perspectives*, edited by Lambros Comitas and David Lowenthal, 163–164. Garden City, N.Y.: Anchor Press.

Cooper, Brittney C. 2017. *Beyond Respectability: The Intellectual Thought of Race Women*. Urbana: University of Illinois Press.

Cooper, Carolyn. 1995. *Noises in the Blood: Orality, Gender, and the "Vulgar" Body of Jamaican Popular Culture*. Durham, N.C.: Duke University Press.

Cooper, Carolyn. 2015. "Professing Slackness: Language, Authority, and Power within the Academy and Without." *e-misferica: Caribbean Rasanblaj* (ed. Gina Ulysse) 12 (1). http://hemisphericinstitute.org/hemi/en/emisferica-121-caribbean-rasanblaj

Craig, Alan K. 1969. "Logwood as a Factor in the Settlement of British Honduras." *Caribbean Studies* 9 (1): 53–62.

Craig, Meg. 1991. *Characters and Caricature in Belizean Folklore*. Belize: Angelus Press

Crichlow, Michaeline A., and Patricia Northover. 2009a. *Globalization and the Post-Creole Imagination: Notes on Fleeing the Plantation*. Durham, N.C.: Duke University Press.

Crichlow, Michaeline A., and Patricia Northover. 2009b. "Homing Modern Freedoms Creolization and the Politics of Making Place." *Cultural Dynamics* 21 (3): 283–316.

Crosby, Alfred W. 2003 [1973]. *The Columbian Exchange: Biological and Cultural Cconsequences of 1492, 30th Anniversary Edition*. Westport, Conn.: Praeger.

Crowe, Frederick. 1850. *The Gospel in Central America: Containing a sketch of the country, physical and geographical, historical and political, moral and religious: a history of the Baptist mission in British Honduras, and of the introduction of the Bible into the Spanish American republic of Guatemala*. London: Charles Gilpin.

Cruikshank, Julie. 2005. *Do Glaciers Listen? Local Knowledge, Colonial Encounters, and Social Imagination*. Seattle: University of Washington Press.

Dampier, William. 1927 [1699]. *A New Voyage Around the World*. London: Argonaut Press.

Davidov, Veronica. 2013. *Ecotourism and Cultural Production: An Anthropology of Indigenous Spaces in Ecuador*. London: Palgrave Macmillan.

De La Cadena, Marisol. 2015. *Earth Beings: Ecologies of Practice across Andean Worlds*. Durham, N.C.: Duke University Press.

De Landa, Manuel. 2006. *A New Philosophy of Society: Assemblage Theory and Social Complexity*. London: Continuum.

Decker, Ken. 2005. *The Song of Kriol: A Grammar of the Kriol Language of Belize*. Belize City: Belize Kriol Project.

Deloria, Vine Jr. 1973. *God Is Red: A Native View of Religion*. Penguin.

Deloria, Vine Jr. 1997. *Red Earth, White Lies: Native Americans and the Myth of Scientific Fact*. Golden, Colo.: Fulcrum Press.

Deleuze, Gilles, and Felix Guattari. 1987. *A Thousand Plateaus: Capitalism and Schizophrenia*. Translated by Brian Massumi. Minneapolis: University of Minnesota Press.

DeLoughrey, Elizabeth M. 2011. "Yam, Roots, and Rot: Allegories of the Provision Grounds." *Small Axe* 15 (1): 58–75.

DeLuca, Kevin, and Anne Demo. 2001. "Imagining Nature and Erasing Class and Race: Carleton Watkins, John Muir, and the Construction of Wilderness." *Environmental History* 6 (4): 541–560.

Descola, Philippe. 2013a. *The Ecology of Others*. Chicago: Prickly Paradigm.

Descola, Philippe. 2013b. *Beyond Nature and Culture*. Chicago: University of Chicago Press.

Descola, Philippe, and Gísli Pálsson, eds. 1996. *Nature and Society: Anthropological Perspectives*. London: Taylor and Francis.

Deshler, William O. 1978. *Proposals for Wildlife Protection and National Parks System Legislation and the Establishment of National Parks and Reserves*. Belize: UNDP/FAO Project Working Document, Forestry Department, Belize.

Doane, Ashley W., and Eduardo Bonilla-Silva. 2003. *White Out: The Continuing Significance of Racism*. New York: Routledge.

Doane, Molly. 2012. *Stealing Shining Rivers: Agrarian Conflict, Market Logic and Conservation in a Mexican Forest*. Tucson: University of Arizona Press.

Doane, Molly. 2014. "From Community Conservation to the Lone (Forest) Ranger: Accumulation by Conservation in a Mexican Forest." *Conservation and Society* 12 (3): 233–244.

Dobson, Narda. 1973. *A History of Belize*. London: Longman Caribbean.

Dominy, Michèle D. 2001. *Calling the Station Home: Place and Identity in New Zealand's High Country*. Lanham, Md.: Rowman and Littlefield.

Douglas, Mary. 1988. *Purity and Danger: An Analysis of the Concepts of Pollution and Taboo*. London: Ark Paperbacks.

Douglass, Lisa. 1992. *The Power of Sentiment: Love, Hierarchy, and the Jamaican Family Elite*. Boulder, Colo.: Westview Press.

Dove, Michael R. 1992. "The Dialectical History of "Jungle" in Pakistan: An Examination of the Relationship between Nature and Culture." *Journal of Anthropological Research* 48 (3): 231–253.

Duffy, Rosaleen. 2013. *A Trip Too Far: Ecotourism, Politics, and Exploitation*. London: Earthscan.

Dundon, Alison. 2005. "The Sense of Sago: Motherhood and Migration in Papua New Guinea and Australia." *Journal of Intercultural Studies* 26 (1–2): 21–37.

Dunlap, Thomas R. 1999. *Nature and the English Diaspora: Environment and History in the United States, Canada, Australia, and New Zealand*. Cambridge: Cambridge University Press.

Dunning, Nicholas P., Sheryl Luzzadder-Beach, Timothy Beach, John G. Jones, Vernon Scarborough, and T. Patrick Culbert. 2002. "Arising from the Bajos: The Evolution of a Neotropical Landscape and the Rise of Maya Civilization." *Annals of the Association of American Geographers* 92 (2): 267–283.

Erazo, Juliet S. 2013. *Governing Indigenous Territories: Enacting Sovereignty in the Ecuadorian Amazon*. Durham, N.C.: Duke University Press.

Escobar, Arturo. 2008. *Territories of Difference: Place, Movements, Life,* Redes. Durham, N.C.: Duke University Press.

Escolar, Diego. 2012 "Boundaries of Anthropology: Empirics and Ontological Relativism in a Field Experience with Anomalous Luminous Entities in Argentina." *Anthropology and Humanism* 37 (1): 27–44.

Escure, Geneviève. 2013. "Belizean Creole." In *The Survey of Pidgin and Creole Languages*. Vol. 1, edited by Susanne Michaelis, Philippe Maurer, Martin Haspelmath, Magnus Huber, 92–100. Oxford: Oxford University Press.

Esselman, Peter C., Juan J. Schmitter-Soto, and J. David Allan. 2013. "Spatiotemporal Dynamics of the Spread of African Tilapias (Pisces: Oreochromis spp) into Rivers of Northeastern Mesoamerica." *Biological Invasion* 15: 1471–1491.

Fabian, Johannes. 2014 [1983]. *Time and the Other: How Anthropology Makes Its Object.* New York: Columbia University Press.

Fanon, Frantz. 2008 [1952]. *Black Skin, White Masks.* New York: Grove Press.

Faria, Caroline, and Sharlene Mollett. 2016. "Critical Feminist Reflexivity and the Politics of Whiteness in the Field." *Gender, Place and Culture* 23 (1): 79–93.

Feagin, Joe. 2009. *The White Racial Frame: Centuries of Racial Framing and Counter Framing.* London: Routledge

Finney, Carolyn. 2014a. "Brave New World? Ruminations on Race in the Twenty-first Century." *Antipode.* 46 (5): 1277–1284.

Finney, Carolyn. 2014b. *Black Faces, White Spaces: Reimagining the Relationship of African Americans to the Great Outdoors.* Durham, N.C.: University of North Carolina Press

Fletcher, Robert. 2014. *Romancing the Wild: Cultural Dimensions of Ecotourism.* Durham, N.C.: Duke University Press.

Fortwangler, Crystal. 2009. "A Place for the Donkey: Natives and Aliens in the US Virgin Islands." *Landscape Research* 34 (2): 205–222.

Fortwangler, Crystal. 2013. "Untangling Introduced and Invasive Animals." *Environment and Society: Advances in Research* 4 (1): 41–59.

Foster, R. J., B. J. Harmsen, D. W. Macdonald, J. Collins, Y. Urbina, R. Garcia, and C. P. Doncaster. 2016. "Wild Meat: A Shared Resource Amongst People and Predators." *Oryx* 50 (1): 63–75.

Fowler, Henry. 1879. *A Narrative of a Journey Across the Unexplored Portion of British Honduras: with a short sketch of the history and resources of the colony.* Belize: Government Press.

Frazier, John W., Florence M. Margai, and Eugene Tettey-Fio. 2003. *Race and Place: Equity Issues in Urban America.* Boulder, Colo.: Westview Press.

Garland, Elizabeth. 2008 "The Elephant in the Room: Confronting the Colonial Character of Wildlife Conservation in Africa." *African Studies Review* 51 (3): 51–74.

Gibson, Katherine, Deborah Bird Rose, and Ruth Fincher, eds. 2015. *Manifesto for Living in the Anthropocene.* New York: Punctum Books.

Gibson-Graham, J. K. 2006a. *The End of Capitalism (as We Knew It): A Feminist Critique of Political Economy.* Minneapolis: University of Minnesota Press.

Gibson-Graham, J. K. 2006b. *A Postcapitalist Politics.* Minneapolis: University of Minnesota Press.

Gibson-Graham, J. K. 2008. "Diverse Economies: Performative Practices for 'Other Worlds.'" *Progress in Human Geography* 32 (5): 613–632.

Gibson-Graham, J. K. 2011. "A Feminist Project of Belonging for the Anthropocene." *Gender, Place and Culture* 18 (1): 1–21.

Gillett, Vincent and George Myvette. 2008. *Vulnerability and Adaptation Assessment of the Fisheries and Aquaculture Industries to Climate Change.* Final Report for the Second National Communication Project. Ministry of Natural Resources and the Environment and UNDP Global Environmental Facility. Belize City.

Gilmore, Ruth Wilson. 2002. "Fatal Couplings of Power and Difference: Notes on Racism and Geography." *Professional Geographer* 54 (1):15–24.

Glave, Dianne D. 2010. *Rooted in the Earth: Reclaiming the African-American Environmental Heritage.* Chicago: Lawrence Hill Books.

Glave, Dianne D., and Mark Stoll. 2006. *To Love the Wind and the Rain: African Americans and Environmental History.* Pittsburgh, Pa.: University of Pittsburgh Press.

Glenn, Evelyn Nakano, ed. 2009. *Shades of Difference: Transnational Perspectives on How and Why Skin Color Matters.* Palo Alto, Calif.: Stanford University Press.

Goldberg, David Theo. 2001. *The Racial State.* Hoboken, N.J.: Blackwell.

Graham, Elizabeth, David M. Pendergast, and Grant D. Jones. 1989. "On the Fringes of Conquest: Maya-Spanish Contact in Colonial Belize." *Science* 246: 1254–1259.

Grandia, Liza. 2012. *Enclosed: Conservation, Cattle, and Commerce Among the Q'eqchi' Maya Lowlanders.* Seattle: University of Washington Press.

Grant, Cedric H. 1976. *The Making of Modern Belize: Politics, Society and British Colonialism in Central America.* Cambridge: Cambridge University Press.

Gravlee, Clarence C. 2009. "How Race Becomes Biology: Embodiment of Social Inequality." *American Journal of Physical Anthropology* 139 (1): 47–57.

Great Belize Productions Channel 5 News. 2003. "Once Feared, Tilapia Now Welcomed in Crooked Tree." March 24, 2003. Accessed November 7, 2015. http://edition.channel5belize.com/archives/15447.

Great Belize Productions Channel 5 News. 2012. "Victim of Crocodile Attack Wakes Up from Induced Coma." July 23, 2012. Accessed October 1, 2017. http://edition.channel5belize.com/archives/73508.

Great Belize Productions Channel 5 News. 2016. "Crooked Tree Fishermen Halted on Haul Day!" June 9, 2016. Accessed October 1, 2017. http://edition.channel5belize.com/archives/130297.

Griffith, David, and Manuel Valdés Pizzini. 2002. *Fishers at Work, Workers at Sea: A Puerto Rican Journey through Labor and Refuge.* Philadelphia: Temple University Press.

Guerron-Montero, C. 2005. "Marine Protected Areas in Panama: Grassroots Activism and Advocacy." *Human Organization* 64 (4): 360–373.

Haenn, Nora. 1999. "The Power of Environmental Knowledge: Ethnoecology and Environmental Conflicts in Mexican Conservation." *Human Ecology* 27 (3): 477–491.

Haenn, Nora. 2016. "Extensionists: A New Elite in Mexican Conservation? Social Dramas, Blurred Identity Boundaries, and Their Environmental Consequences in Mexican Conservation." *Current Anthropology* 57 (2): 197–218.

Hannam, Kevin. 2007. "Shooting Tigers as Leisure in Colonial India," In *Tourism and the Consumption of Wildlife: Hunting, Shooting and Sport Fishing,* edited by Brent Lovelock, 100–110. London: Routledge.

Haraway, Donna. 1991. *Simians, Cyborgs, and Women: The Reinvention of Nature.* New York: Routledge.

Haraway, Donna. 2003. *The Companion Species Manifesto: Dogs, People, and Significant Otherness.* Chicago: Prickly Paradigm Press.

Haraway, Donna. 2015. "Anthropocene, Capitalocene, Plantationocene, Chthulucene: Making Kin." *Environmental Humanities* 6: 159–165.

Harrison, Faye Venetia, ed. 1991. *Decolonizing Anthropology Moving Further toward an Anthropology for Liberation.* Arlington, Va.: Association of Black Anthropologists, American Anthropological Association.

Harrison-Buck, Eleanor. 2014. "Ancient Maya Wetland Use in the Eastern Belize Watershed." *Research Reports in Belizean Archaeology* 11: 245–258.

Hartshorn, Gary et al. 1984. *Belize, Country Environmental Profile: A Field Study.* Belize City: Robert Nicolait.

Hathaway, Michael J. 2013. *Environmental Winds: Making the Global in Southwest China*. Berkeley: University of California Press.

Hathaway, Michael J. Forthcoming. "Making More than Human Worlds: Mushrooms in the Global Economy."

Henderson, Captain George. 1809. *An Account of the British Settlement of Honduras; Being a Brief View of its Commercial and Agricultural Resources, Soil, Climate, Natural History, Etc. To Which Are Added, Sketches of the Manners and Customs of the Mosquito Indians, Preceded by the Journal of a Voyage to the Mosquito Shore. Illustrated by a Map.* London: C. and R. Baldwin.

Hoelle, Jeffrey. 2015. *Rainforest Cowboys: The Rise of Ranching and Cattle Culture in Western Amazonia*. Austin: University of Texas Press.

Hoffman, Kelly M., Sophie Trawalter, Jordan R. Axt, and M. Norman Oliver. 2016. "Racial Bias in Pain Assessment and Treatment Recommendations, and False Beliefs about Biological Differences between Blacks and Whites." *Proceedings of the National Academy of Sciences* 113 (16): 4296–4301.

Hoffman, Odile. 2014. *British Honduras: The Invention of a Colonial Territory: Mapping and Spatial Knowledge in the 19th Century.* Benque Viejo del Carmen: Cubola Productions.

The Honduras Almanack for the year of our Lord 1830. Belize: Legislative Assembly.

The Honduras Almanack for the year of our Lord 1839. Belize: Legislative Assembly.

hooks, bell. 1990. "Choosing the Margin as a Space of Radical Openness" In *Race, Gender and Cultural Politics*, 145–153. Boston: South End Press.

HoSang, Daniel, Oneka LaBennett, and Laura Pulido. 2012. *Racial Formation in the Twenty-First Century.* Berkeley: University of California Press.

Hughey, Matthew W. 2010. "The (Dis)Similarities of White Racial Identities: The Conceptual Framework of 'Hegemonic' Whiteness." *Ethnic and Racial Studies* 33 (8): 1289–1309.

Hummel, Cornelius. 1921. *Report on the Forests of British Honduras with Suggestions for a Far Reaching Forest Policy.* London: Her Majesty's Stationary Office.

Igoe, Jim, and Dan Brockington. 2007. "Neoliberal Conservation: A Brief Introduction." *Conservation and Society* 5 (4): 432–449.

Ingold, Tim. 2000. *The Perception of the Environment: Essays on Livelihood, Dwelling and Skill.* London: Routledge.

Jaffe, Rivke. 2008. "A View from the Concrete Jungle: Diverging Environmentalisms in the Urban Caribbean." *New West Indian Guide/Nieuwe West-Indische Gids* 80 (3–4): 221–243.

Jaffe, Rivke. 2016. *Concrete Jungles: Urban Pollution and the Politics of Difference in the Caribbean*. London: Oxford University Press.

Jahoda, Gustav. 1999. *Images of Savages: Ancient Roots of Modern Prejudice in Western Culture*. London: Routledge.

Johnson, Jay T., Garth Cant, Richard Howitt, and Evelyn Peters. 2007. "Creating Anticolonial Geographies: Embracing Indigenous Peoples? Knowledges and Rights." *Geographical Research*. 45 (2): 117–120.

Johnson, Melissa. 2003. "The Making of Race and Place in Nineteenth-Century British Honduras." *Environmental History* 8 (4): 598–617.

Johnson, Melissa. 2005. "Racing Nature and Naturalizing Race: Rethinking the Nature of Creole and Garifuna Communities." *Belizean Studies* 27 (2): 43–56.

Johnson, Melissa. 2015. "Creolized Conservation: A Belizean Creole Community Encounters a Wildlife Sanctuary." *Anthropological Quarterly* 88 (1): 67–95.

Jones, Grant D. 1989. *Maya Resistance to Spanish Rule: Time and History on a Colonial Frontier.* Albuquerque: University of New Mexico Press.

Joseph, Gilbert. 1987. "The Logwood Trade and Its Settlements." In *Studies in Belizean History,* edited by Lita Hunter Krohn, 39–42. Belize City: St. Johns College.

Joseph, Gilbert. 1989. "John Coxon and the Role of Buccaneering in the Settlement of the Yucatan Colonial Frontier." *Belizean Studies* 17 (3): 1–21.

Joseph, Tiffany D. 2015. *Race on the Move: Brazilian Migrants and the Global Reconstruction of Race.* Palo Alto, Calif.: Stanford University Press.

Judd, Karen. 1990. "Who Will Define Us? Creolization in Belize." *Second Annual Studies on Belize Conference,* 29–41. Belize City: SPEAR.

Judd, Karen. 1992. "Elite Reproduction and Ethnic Identity in Belize." Ph.D. diss., Graduate Center of the City University of New York.

Kamugisha, Aaron. 2016. "The Black Experience of New World Coloniality." *Small Axe* 20 (1): 129–145.

Kane, Stephanie. 1994. *The Phantom Gringo Boat: Shamanic Discourse and Development in Panama.* Washington, D.C.: Smithsonian Institution Press.

Kang, Tingyu. 2009. "Homeland Re-Territorialized: Revisiting the Role of Geographical Places in the Formation of Diasporic Identity in the Digital Age." *Information, Communication and Society* 12 (3): 326–343.

Kawa, Nicholas. 2016. *Amazonia in the Anthropocene: People, Soils, Plants, Forests.* Austin: University of Texas Press.

Kelly, John D. 2014. "Introduction: The Ontological Turn in French Philosophical Anthropology." *HAU: Journal of Ethnographic Theory* 4 (1): 259–269.

Kerns, Virginia. 1983. *Women and the Ancestors: Black Carib Kinship and Ritual.* Urbana: University of Illinois Press.

Khan, Aisha. 1997 "Rurality and 'Racial' Landscapes in Trinidad." In *Knowing Your Place: Rural Identity and Cultural Hierarchy,* edited by Barbara Ching and Gerald Creed, 39–69. New York: Routledge.

Kim, Claire Jean. 2015. *Dangerous Crossings: Race, Species, and Nature in a Multicultural Age.* New York: Cambridge University Press

King, Russell, and Anastasia Christou. 2011. "Of Counter-Diaspora and Reverse Transnationalism: Return Mobilities to and from the Ancestral Homeland." *Mobilities* 6 (4): 451–466.

Knight, John, ed. 2000. *Natural Enemies: People-wildlife Conflicts in Anthropological Perspective.* London: Routledge.

Knudsen, Ståle. 2014 "Multiple Sea Snails: The Uncertain Becoming of an Alien Species." *Anthropological Quarterly* 87 (1): 59–91.

Komito, Lee. 2011. "Social Media and Migration: Virtual Community 2.0." *Journal of the American Society for Information Science and Technology* 62 (6): 1075–1086.

Kosek, Jake. 2006. *Understories: The Political Life of Forests in Northern New Mexico.* Durham, N.C.: Duke University Press.

Lanham, J. Drew. 2016. *The Home Place: Memoirs of a Colored Man's Love Affair with Nature*. Minneapolis, Minn.: Milkweed Press.

Latour, Bruno. 2005. *Reassembling the Social: An Introduction to Actor-Network-Theory*. London: Oxford University Press.

Leslie, Robert. 1995. *A History of Belize: A Nation in the Making*. Benque Viejo del Carmen, Belize: Cubola Productions.

Leslie, Vernon. 1987. "The Belize River Boat Traffic." *Caribbean Quarterly* 33 (3 and 4): 1–28.

Li, Fabiana. 2015. *Unearthing Conflict: Corporate Mining, Activism, and Expertise in Peru*. Durham, N.C.: Duke University Press.

Linebaugh, Peter, and Marcus Rediker. 2013 [2001]. *The Many-Headed Hydra: Sailors, Slaves, Commoners, and the Hidden History of the Revolutionary Atlantic*. Boston: Beacon Press.

Lipsitz, George. 2007. "The Racialization of Space and the Spatialization of Race: Theorizing the Hidden Architecture of Landscape." *Landscape Journal* 26 (1): 10–23.

Lipsitz, George. 2011. *How Racism Takes Place*. Philadelphia: Temple University Press.

Lloyd, Kate, Sandie Suchet-Pearson, Sarah Wright, and Lak Lak Burarrwanga. 2010. "Stories of Crossings and Connections from Bawaka, North East Arnhem Land, Australia." *Social and Cultural Geography* 11 (7): 701–717.

Lowenthal, David. 1961. "Caribbean Views of Caribbean Land." *Canadian Geographer* 5 (2): 1–9.

Lucero, L. J., S. L. Fedick, N. P. Dunning, D. L. Lentz, and V. L. Scarborough. 2014. "Chapter 3 Water and Landscape: Ancient Maya Settlement Decisions." *Archeological Papers of the American Anthropological Association* 24: 30–42.

MacKenzie, John. 1997. *The Empire of Nature: Hunting, Conservation and British Imperialism*. Manchester: Manchester University Press.

Mackler, Robert, and Osmany Salas. 1994. *Management Plan: Crooked Tree Wildlife Sanctuary*. Belize City: Belize Audubon Society.

Malkki, Liisa. 1992. "National Geographic: The Rooting of Peoples and the Territorialization of National Identity among Scholars and Refugees." *Cultural Anthropology* 7 (1): 24–44.

Marks, Jonathan. 1995. *Human Biodiversity: Genes, Race, and History*. New York: Aldine de Gruyter.

Marks, Jonathan. 2017. *Is Science Racist?* Malden, Mass.: Polity Press.

McAllister, Elizabeth. 2002. *Rara: Vodou, Power, and Performance in Haiti and Its Diaspora*. Berkeley: University of California Press.

McKittrick, Katherine. 2006. *Demonic Grounds: Black Women and the Cartographies of Struggle*. Minneapolis: University of Minnesota Press.

McKittrick, Katherine, ed. 2015. *Sylvia Wynter: On Being Human as Praxis*. Durham, N.C.: Duke University Press.

McKittrick, Katherine, and Clyde Woods, eds. 2007. *Black Geographies and the Politics of Place*. Toronto: Between the Lines Press.

McNeill, John Robert. 2010. *Mosquito Empires: Ecology and War in the Greater Caribbean, 1620–1914*. Cambridge: Cambridge University Press.

Medina, Laurie Kroshus. 2003. "Commoditizing Culture: Tourism and Maya Identity." *Annals of Tourism Research: A Social Sciences Journal* 30 (2): 353–368.

Medina, Laurie Kroshus. 2004. *Negotiating Economic Development: Identity Formation and Collective Action in Belize*. Tucson: University of Arizona Press

Medina, Laurie Kroshus. 2012. "The Uses of Ecotourism Articulating Conservation and Development Agendas in Belize." In *Global Tourism: Cultural Heritage and Economic Encounters*, edited by Sarah Lyons and E. Christian Wells, 227–250. Lanham, Md.: Alta Mira Press.

Medina, Laurie Kroshus. 2015. "Governing through the Market: Neoliberal Environmental Government in Belize." *American Anthropologist* 117 (2): 272–284.

Medina, Laurie Kroshus. 2017. Personal communication.

Meerman, Jan C. 2005. "Compilation of Information on Biodiversity in Belize." *Report to INBio and the Chief Forest Officer*, Forest Department of the Ministry of Natural Resources and the Environment, Belize.

Merchant, Carolyn. 1989. *The Death of Nature: Women, Ecology, and the Scientific Revolution*. New York: Harper and Row.

Merchant, Carolyn. 2003. "Shades of Darkness: Race and Environmental History." *Environmental History* 8 (3): 380–394.

Mikulak, Marcia L. 2011. "The Symbolic Power of Color: Constructions of Race, Skin-Color, and Identity in Brazil." *Humanity and Society* 35 (1–2): 62–99.

Mills, Charles W. 1997. *The Racial Contract*. Ithaca, N.Y.: Cornell University Press.

Mills, Charles W. 2011. "Global White Supremacy." In *White privilege: Essential Readings on the Other Side of Racism*, edited by Paula S. Rothenberg, 95–102. New York: Worth.

Mintz, Sidney W. 1974. *Caribbean Transformations*. New York: Aldine.

Mintz, Sidney W. 1986. *Sweetness and Power: The Place of Sugar in Modern History*. New York: Penguin.

Mintz, Sidney W. 1996. *Tasting Food, Tasting Freedom: Excursions into Eating, Culture, and the Past*. Boston: Beacon Press.

Mintz, Sidney W. 2010. "Houses and Yards among Caribbean Peasantries." In *Perspectives on the Caribbean: A Reader in Culture, History, and Representation*, edited by Philip W. Scher, 10–24. Chichester: Wiley-Blackwell.

Mintz, Sidney W., and Richard Price. 1992. *The Birth of African-American Culture: An Anthropological Approach*. Boston: Beacon Press.

Minute Paper 4056 of 1915. National Archives of Belize, Belmopan.

Minute Paper 632 of 1951. National Archives of Belize, Belmopan.

Minute Paper 1016 of 1951. National Archives of Belize, Belmopan.

Mitchell, Brent A., Zoe Walker, and Paul Walker. 2017. "A Governance Spectrum: Protected Areas in Belize." *Parks* 23 (1): 45–60.

Moberg, Mark. 1997. *Myths of Ethnicity and Nation: Immigration, Work, and Identity in the Belize Banana Industry*. Knoxville: University of Tennessee Press.

Momsen, Janet H., ed. 1993. *Women and Change in the Caribbean: A Pan-Caribbean Perspective*. Kingston, Jamaica: Ian Randle.

Moore, Amelia. 2012. "The Aquatic Invaders: Marine Management Figuring Fishermen, Fisheries, and Lionfish in the Bahamas." *Cultural Anthropology* 27 (4): 667–688.

Moore, Amelia. 2016. "Anthropocene Anthropology: Reconceptualizing Contemporary Global Change." *Journal of the Royal Anthropological Institute* 22 (1): 27–46.

Moore, Donald S., Jake Kosek, and Anand Pandian, eds. 2003. *Race, Nature, and the Politics of Difference.* Durham, N.C.: Duke University Press.

Morgan, George, Cristina Rocha, and Scott Poynting. 2005. "Grafting Cultures: Longing and Belonging in Immigrants' Gardens and Backyards in Fairfield." *Journal of Intercultural Studies* 26 (1/2): 93–105.

Morris, Daniel. 1883. *The Colony of British Honduras, Its Resources and Prospects.* London: E. Stanford.

Mowforth, Martin, and Ian Munt. 2003. *Tourism and Sustainability: Development, Globalisation and New Tourism in the Third World.* London: Routledge.

Munro, David. 1983. "The Conservation of Nature in Belize." In *Resources and Development in Belize,* edited by G. M. Robinson and P. A. Furley, 139–156. Edinburgh: Department of Geography, University of Edinburgh.

Munt, Ian. 1994. "Eco-tourism or Ego-tourism." *Race and Class* 36 (1): 49–60.

Nayak, Anoop. 2006. "After Race: Ethnography, Race and Post-Race Theory." *Ethnic and Racial Studies* 29 (3): 411–430.

Naylor, Robert A. 1989. *Penny Ante Imperialism: The Mosquito Shore and the Bay of Honduras, 1600–1914: A Case Study in British Informal Empire.* Madison, N.J.: Fairleigh Dickinson University Press.

Nelson, Richard K. 1986 [1973]. *Hunters of the Northern Forest: Designs for Survival among the Alaskan Kutchin.* Chicago: University of Chicago Press.

Nesbitt, Nick. 2013. "Pre-face: Escaping Race." In *Deleuze and Race,* edited by Arun Saldanha and Jason Michael Adams, 1–5. Edinburgh: Edinburgh University Press.

Nixon, Rob. 2011. *Slow Violence and the Environmentalism of the Poor.* Cambridge, Mass.: Harvard University Press.

Norwood, Kimberly Jade. 2014. *Color Matters: Skin Tone Bias and the Myth of a Post-Racial America.* New York: Routledge.

Nuenninghoff, Sybille, Michele H. Lemay, Cassandra Rogers, and Dougal Martin. 2014. *Sustainable Tourism in Belize.* Belize City: Inter-American Development Bank.

NurMuhammad, Rizwangul, Heather A. Horst, Evangelina Papoutsaki, and Giles Dodson. 2016. "Uyghur Transnational Identity on Facebook: On the Development of a Young Diaspora." *Identities* 23 (4): 485–499.

Nyamnjoh, Francis B., and Ben Page. 2002. "*Whiteman Kontri* and the Enduring Allure of Modernity Among Cameroon Youth." *African Affairs* 101:607–634.

Ogden, Laura. 2011. *Swamplife: People, Gators, and Mangroves Entangled in the Everglades.* Minneapolis: University of Minnesota Press.

Ogden, Laura A., Billy Hall, and Kimiko Tanita. 2013. "Animals, Plants, People, and Things: A Review of Multispecies Ethnography." *Environment and Society: Advances in Research.* 4 (1): 5–24.

Old Records, 1867, R82, folio 892–908. National Archives of Belize, Belmopan.

Olwig, Karen Fog. 2002 [1992]. *Global Culture, Island identity: Continuity and Change in the Afro-Caribbean Community of Nevis.* London: Routledge.

Omi, Michael, and Howard Winant. 1986. *Racial Formation in the United States: From the 1960s to 1980s.* New York: Routledge.

Outka, Paul. 2008. *Race and Nature from Transcendentalism to the Harlem Renaissance.* New York: Palgrave Macmillan.

Palacio, Joseph Orlando, ed. 2006. *The Garifuna: A Nation across Borders: Essays in Social Anthropology.* Benque Carmen de Vieja, Belize: Cubola Productions.

Pearce, Fred. 2015. *The New Wild: Why Invasive Species Will Be Nature's Salvation*. Boston: Beacon Press.

Pilgrim, Sarah, and Jules N. Pretty. 2010. *Nature and Culture: Rebuilding Lost Connections*. London: Earthscan.

Platt, Elizabeth. 1998. "Forest Management and Conservation in Belize: A Brief Background." In *Timber, Tourists, and Temples: Conservation and Development in the Maya Forest of Belize, Guatemala, and Mexico*, edited by Richard B Primack, David Bray, Hugo Galletti, and Ismael Ponciano, 125–135. Washington, D.C.: Island Press.

Platt, Steven G., Thomas R. Rainwater, John B. Thorbjarnarson, and Scott T. McMurry. 2008. "Reproductive Dynamics of a Tropical Freshwater Crocodilian: Morelet's Crocodile in Northern Belize." *Journal of Zoology* 275 (2): 177–189.

Platt, Steven G., Luis Sigler, and Thomas R. Rainwater. 2010. "Morelet's Crocodile Crocodylus moreletii." In *Crocodiles. Status Survey and Conservation Action Plan*, 3rd ed., edited by S.C. Manolis and C. Stevenson. Crocodile Specialist Group: Darwin, 79–83. Gland, Switzerland: IUCN.

Platt, Steven G., and John B. Thorbjarnarson. 2000."Population Status and Conservation of Morelet's Crocodile, *Crocodylus moreletii*, in Northern Belize." *Biological Conservation* 96 (1): 21–29.

Plumwood, Val. 2012. *The Eye of the Crocodile*. Canberra: Australian National University Press.

Polisar, John, and Robert H. Horwich 1994. "Conservation of the Large, Economically Important River Turtle *Dermatemys mawii* in Belize." *Conservation Biology* 8 (2): 338–340.

Pulido, Laura. 2017. "Geographies of Race and Ethnicity: Environmental Racism, Racial Capitalism and State-Sanctioned Violence." *Progress in Human Geography* 41 (4): 524–533.

Pyburn, K. Anne. 2003. "The Hydrology of Chau Hiix." *Ancient Mesoamerica* 14 (1): 123–129.

Quammen, David. 2004. *Monster of God: The Man-Eating Predator in the Jungles of History and the Mind*. New York: Norton.

Quijano, Anibal. 2000. "Coloniality of Power: Eurocentrism and Latin America." *Neplanta: Views from the South* 1 (3): 533–580.

Quijano, Jose Antonio Calderon. 1944. *Belice 1663?–1821: Historia de los Establecimientos Britanicos del Rio Valis hasta la Independencia de Hispanoamerica*. Publicaciones de la Escuela de Estudios Hispano-Americanos de la Universidad de Sevilla, V 9.

Rabinowitz, Alan. 2000 [1987]. *Jaguar: One Man's Struggle to Establish the World's First Jaguar Preserve*. Washington, D.C.: Island Press.

Rainwater, Thomas R., Thomas Pop, Octavio Cal, Anthony Garel, Steven G. Platt, and Rick Hudson. 2012. "A Recent Countrywide Status Survey of the Critically Endangered Central American River Turtle (*Dermatemys mawii*) in Belize." *Chelonian Conservation and Biology* 11 (1): 97–107.

Razack, Sherene. 2002. *Race, Space and the Law: Unmapping a White Settler Society*. Toronto: Between the Lines Press.

Redford, Kent. 1991. "The Ecologically Noble Savage." *Cultural Survival Quarterly* 15 (1): 46–48.

Restrepo, Eduardo, and Arturo Escobar. 2005. "'Other Anthropologies and Anthropologies Otherwise': Steps to a World Anthropology Framework." *Critique of Anthropology* 25 (2): 99–129.

Richardson, David M., ed. 2011. *Fifty Years of Invasion Ecology: The Legacy of Charles Elton*. Chichester: Wiley-Blackwell.

Ricourt, Milagros. 2016. *The Dominican Racial Imaginary: Surveying the Landscape of Race and Nation in Hispaniola*. New Brunswick, N.J.: Rutgers University Press.

Robbins, Paul. 2007. *Lawn People: How Grasses, Weeds, and Chemicals Make Us Who We Are*. Philadelphia: Temple University Press.

Roberson, Beth. 2015. "Jaguar Predation: Ranchers and Conservationists Strive Together for Answers Toledo's Ya'axche Hosts Experts from Panthera." *Belize Agricultural Report*, August 1, Issue 29.

Roberts, Dorothy E. 2011. *Fatal Invention: How Science, Politics, and Big Business Recreate Race in the Twenty-First Century*. New York: New Press.

Roberts, Neil. 2014. "Violence, Livity, Freedom." *Small Axe* 18 (1): 181–192.

Roberts, Neil. 2015. *Freedom as Marronage*. Chicago: University of Chicago Press.

Roelvink, Gerda. 2015. "Learning to Be Affected by Earth Others." In *Manifesto for Living in the Anthropocene*, edited by Katherine Gibson, Deborah Bird Rose, and Ruth Fincher, 57–62. Brooklyn, N.Y.: Punctum Books.

Roland, L. Kaifa. 2011. *Cuban Color in Tourism and La Lucha: An Ethnography of Racial Meanings*. New York: Oxford University Press.

Rose, Deborah Bird. 2004. *Reports from a Wild Country: Ethics for Decolonisation*. Sydney: University of New South Wales Press.

Ross, Kirstie. 2008. *Going Bush: New Zealanders and Nature in the Twentieth Century*. Auckland: Auckland University Press.

Rubenstein, Hymie. 1987. *Coping with Poverty: Adaptive Strategies in a Caribbean Village*. London: Routledge.

Salazar, Noel B. 2011. "The Power of Imagination in Transnational Mobilities." *Identities: Global Studies in Culture and Power* 18 (6): 576–598.

Saldanha, Arun. 2006. "Reontologising Race: The Machinic Geography of Phenotype." *Environment and Planning D: Society and Space* 24 (1): 9–24.

Saldanha, Arun. 2007. *Psychedelic White: Goa Trance and the Viscosity of Race*. Minneapolis: University of Minnesota Press.

Saldanha, Arun. 2012. "Assemblage, Materiality, Race, Capital." *Dialogues in Human Geography* 2 (2): 194–97.

Saldanha, Arun, and Jason Michael Adams, eds. 2013. *Deleuze and Race*. Edinburgh: Edinburgh University Press.

Salmon, William. 2014. "The Contrastive Discourse Marker ata in Belizean Kriol." *Lingua* 143: 86–102.

Salmon, William. 2015. "Language Ideology, Gender and Varieties of Belizean Kriol." *Journal of Black Studies* 46 (6): 605–625.

Salmon, William, and Jennifer Gómez Menjívar. 2014. "Whose Kriol Is Moa Beta? Prestige and Dialects of Kriol in Belize." *Annual Meeting of the Berkeley Linguistics Society* 40: 456–469.

Salmon, William, and Jennifer Gómez Menjívar. 2016 "Language Variation and Dimensions of Prestige in Belizean Kriol." *Journal of Pidgin and Creole Languages* 31 (2): 316–360.

Santiago-Irizarry, Vilma. 2008. “Transnationalism and Migration: Locating Sociocultural Practices among Mexican Immigrants in the United States.” *Reviews in Anthropology* 37 (1): 16–40.

Sendejo, Brenda. 2014. “Methodologies of the Spirit: Reclaiming our Lady of Guadalupe and Discovering Tonantzin within and beyond the *Nepantla* of Academia.” In *Fleshing the Spirit: Spirituality and Activism in Chicana, Latina, and Indigenous Women's Lives*, edited by Elisa Facio and Irene Lara, 81–101. Tucson: University of Arizona Press.

Sheller, Mimi. 2003a. *Consuming the Caribbean: From Arawaks to Zombies*. New York: Routledge.

Sheller, Mimi. 2003b. “Creolization in Discourses of Global Culture.” In *Uprootings/Regroundings: Questions of Home and Migration*, edited by Sara Ahmed, Claudia Castañeda, Anne-Marie Fortie, and Mimi Sheller, 273–294. London: Berg.

Shillington, Laura. 2008. “Being(s) in Relation at Home: Socio-natures of Patio ‘Gardens’ in Managua, Nicaragua.” *Social and Cultural Geography* (9) 7: 755–776

Shoman, Assad. 1994. *Thirteen Chapters of a History of Belize*. Belize City: Angelus Press.

Simmons, Kimberly E. 2009. *Reconstructing Racial Identity and the African Past in the Dominican Republic*. Gainesville: University of Florida Press.

Slocum, Karla. 2017. “Caribbean Free Villages: Toward an Anthropology of Blackness, Place, and Freedom.” *American Ethnologist* 44 (3): 425–434.

Slocum, Rachel. 2011. “Race in the Study of Food.” *Progress in Human Geography* 35 (3): 303–327.

Smith, Jennie Marcelle. 2001. *When the Hands Are Many: Community Organization and Social Change in Rural Haiti*. Ithaca, N.Y.: Cornell University Press.

Smith, Linda Tuhiwai. 1999. *Decolonizing Methodologies: Research and Indigenous Peoples*. London: Zed.

Smith, Michael Peter, and Luis E Guarnizo. 1998. “The Location of Transnationalism.” In *Transnationalism from Below*, edited by Smith and Guarnizo, 3–34. New Brunswick, N.J.: Transaction.

Stark, David. 2015. *Slave Families and the Hato Economy in Puerto Rico*. Gainesville: University of Florida Press.

Stavrakis, Olga.1979. “The Effect of Agricultural Change upon Social Relations in a Village in Northern Belize.” Ph.D. diss., University of Minnesota.

Steinberg, Michael K. 2016. “Jaguar Conservation in Southern Belize: Conflicts, Perceptions, and Prospects Among Mayan Hunters.” *Conservation and Society* 14 (1): 13–20.

Stewart, Charles. 2007. *Creolization: History, Ethnography, Theory*. Walnut Creek, Calif.: Left Coast Press.

Stone, Michael. 1994. “Caribbean Nation, Central American State: Ethnicity, Race and National Formation in Belize, 1798–1990.” Ph.D. diss., University of Texas, Austin.

Subramaniam, Banu. 2001. “The Aliens Have Landed! Reflections on the Rhetoric of Biological Invasions.” *Meridians: Feminism, Race, Transnationalism* 2 (1): 26–40.

Subramaniam, Banu. 2014. *Ghost Stories for Darwin: The Science of Variation and the Politics of Diversity*. Urbana-Champaign: University of Illinois Press.

Sundberg, Juanita. 2004. “Identities in the Making: Conservation, Gender and Race in the Maya Biosphere Reserve, Guatemala.” *Gender, Place and Culture* 11 (1): 43–66.

Sundberg, Juanita. 2008a. "Placing Race in Environmental Justice Research in Latin America." *Society and Natural Resources* 21: 569–582.

Sundberg, Juanita. 2008b. "Tracing Race: Mapping Environmental Formations in Environmental Justice Research in Latin America." In *Environmental Justice in Latin America: Problems, Promise, and Practice*, edited by David V. Carruthers, 25–47. Cambridge, Mass.: MIT Press.

Sundberg, Juanita. 2014. "Decolonizing Posthumanist Geographies." *Cultural Geographies* 21 (1): 33–47.

Suzuki, Yuja. 2016. *The Nature of Whiteness: Race, Animals and Nation in Zimbabwe.* Seattle: University of Washington Press.

TallBear, Kim. 2015. "Dossier: Theorizing Queer Inhumanisms: An Indigenous Reflection on Working beyond the Human/Not Human." *GLQ: A Journal of Lesbian and Gay Studies* 21 (2–3): 230–235.

Taylor, Dorceta E. 1997. "American Environmentalism: The Role of Race, Class and Gender in Shaping Activism 1820–1995." *Race, Gender and Class* 3 (1): 16–62.

Taylor, Lawrence. 1981. "'Man the Fisher': Salmon Fishing and the Expression of Community in a Rural Irish Settlement." *American Ethnologist* 8 (4): 774–788.

Templeton, Alan R. 2013. "Biological Races in Humans." *Studies in the History and Philosophy of the Biological and Biomedical Sciences* 44 (3): 262–271.

Thomas, Deborah A. 2004. *Modern Blackness: Nationalism, Globalization, and the Politics of Culture in Jamaica.* Durham, N.C.: Duke University Press.

Thomas, Deborah, and Karla Slocum. 2007. "Introduction: Locality in Today's Global Caribbean: Shifting Economies of Nation, Race and Development." *Identities* 14 (1/2): 1–18.

Tobler, Mathias W., Samia E. Carrillo-Percastegui, Alfonso Zúñiga Hartley, and George V. N. Powell. 2013. "High Jaguar Densities and Large Population Sizes in the Core Habitat of the Southwestern Amazon." *Biological Conservation* 159: 375–381.

Todd, Zoe. 2016. "An Indigenous Feminist's Take on the Ontological Turn: 'Ontology' Is Just Another Word for Colonialism." *Journal of Historical Sociology* 29 (1): 4–22.

Trouillot, Michel-Rolph. 1992. "The Caribbean Region: An Open Frontier in Anthropological Theory." *Annual Review of Anthropology* 21: 19–42.

Trouillot, Michel-Rolph. 2003 [1991]. "Anthropology and the Savage Slot: The Poetics and Politics of Otherness." In *Global Transformations*, 7–28. New York: Palgrave Macmillan.

Tsing, Anna Lowenhaupt. 2015. *The Mushroom at the End of the World: On the Possibility of Life in Capitalist Ruins.* Princeton, N.J.: Princeton University Press.

Tucker, Richard P. 1982. "The Forests of the Western Himalayas: The Legacy of British Colonial Administration." *Forest and Conservation History* 26 (3): 112–123.

Twine, France Winddance, Charles Gallagher, and Ruth Frankenberg. 2008. *Whiteness and White Identities.* Milton Park, Abingdon: Routledge.

Ulysse, Gina A. 1999. "Uptown Ladies and Downtown Women: Female Representations of Class and Color in Jamaica." In *Representations of Blackness and the Performance of Identities*, edited by Jean M. Rahier, 147–172. Santa Barbara, Calif.: Bergin and Garvey.

Ulysse, Gina A. 2002. "Conquering Duppies in Kingston: Miss Tiny and Me, Fieldwork Conflicts, and Being Loved and Rescued." *Anthropology and Humanism Quarterly* 27 (1): 10–26.

Ulysse, Gina A. 2007. *Downtown Ladies: Informal Commercial Importers, a Haitian Anthropologist and Self-Making in Jamaica*. Chicago: University of Chicago Press.

Ulysse, Gina A. 2015. *Why Haiti Needs New Narratives: A Post-Quake Chronicle*. Middletown, Conn.: Wesleyan University Press.

Ulysse, Gina A. Website. www.ginaathenaulysse.com.

Uring, Nathaniel. 1726. *A History of the Voyages and Travels of Captain Nathaniel Uring*. London: Wilkins and Peel.

Urry, John. 1990. *The Tourist Gaze: Leisure and Travel in Contemporary Societies*. London: Sage.

Waight, Lydia, and Judy Lamb. 1999. *The Belize Audubon Society: The First 30 Years*. Producciones de la Hamaca. http://producciones-amaca.com/books/bashistory.html

Wainwright, Joel. 2008. *Decolonizing Development: Colonial Power and the Maya*. Malden, Mass.: Blackwell.

Walcott, Derek. 1992. "Nobel Lecture: The Antilles: Fragments of Epic Memory." http://nobelprize.org/nobel_prizes/literature/laureates/1992/walcott-lecture.html.

Walley, Christine J. 2013. *Exit Zero: Family and Class in Postindustrial Chicago*. Chicago: University of Chicago Press.

Wardi, Anissa Janine. 2011. *Water and African American Memory: An Ecocritical Perspective*. Gainesville: University Press of Florida.

Warren, Jonathan W. 2001. *Racial Revolutions: Antiracism and Indian Resurgence in Brazil*. Durham, N.C.: Duke University Press.

Waterston, Alisse. 2013. *My Father's Wars: Migration, Memory, and the Violence of a Century*. New York: Routledge.

Waterston, Alisse, and Barbara Rylko-Bauer. 2006. "Out of the Shadows of History and Memory: Personal Family Narratives in Ethnographies of Rediscovery." *American Ethnologist* 33 (3): 397–412.

Watson, Annette, and Orville H. Huntington. 2008. "They're Here—I Can Feel Them: The Epistemic Spaces of Indigenous and Western Knowledges." *Social and Cultural Geography* 9 (3): 257–281.

Watson, Annette, and Orville Huntington. 2014. "Transgressions of the Man on the Moon: Climate Change, Indigenous Expertise, and the Posthumanist Ethics of Place and Space." *GeoJournal: Spatially Integrated Social Sciences and Humanities* 79 (6): 721–736.

Watts, Vanessa. 2013. "Indigenous Place-Thought and Agency Amongst Humans and Non Humans (First Woman and Sky Woman Go on a European World Tour!)." *Decolonization: Indigeneity, Education and Society* 2 (1): 20–34.

Weheliye, Alexander G. 2014. *Habeas Viscus: Racializing Assemblages, Biopolitics, and Black Feminist Theories of the Human*. Durham, N.C.: Duke University Press.

Wells, Marilyn McKillop. 2015. *Among the Garifuna: Family Tales and Ethnography from the Caribbean Coast*. Tuscaloosa: University of Alabama Press.

Werry, Margaret. 2008. "Tourism Race and the State of Nature: On the Bio-Poetics of Government." *Cultural Studies* 22 (3–4): 391–411.

Werry, Margaret. 2011. *The Tourist State: Performing Leisure, Liberalism, and Race in New Zealand*. Minneapolis: University of Minnesota Press.

West, Paige. 2006. *Conservation Is Our Government Now: The Politics of Ecology in Papua New Guinea*. Durham, N.C.: Duke University Press.

West, Paige, and James G. Carrier. 2004. "Getting Away from It All? Ecotourism and Authenticity." *Current Anthropology* 45 (4): 483–98.

West, Paige, James Igoe, and Dan Brockington. 2006. "Parks and Peoples: The Social Impact of Protected Areas." *Annual Review of Anthropology* 35: 251–277.

Whatmore, Sarah. 2002. *Hybrid Geographies: Natures, Cultures, Spaces*. London: Sage.

Whatmore, Sarah. 2006. "Materialist Returns: Practising Cultural Geography in and for a More-than-Human World." *Cultural Geographies* 13 (4): 600–609.

White, Richard. 1995. "'Are You an Environmentalist or Do You Work for a Living?' Work and Nature." In *Uncommon Ground: Toward Reinventing Nature*, edited by William Cronon, 171–185. New York: Norton.

Wilk, Richard. 1989. "Colonial Time and T.V. Time: Media and Historical Consciousness in Belize." *Belizean Studies* 17 (1): 3–13.

Wilk, Richard. 1990. "Consumer Goods as Dialogue about Development." *Culture and History* 7: 79–100

Wilk, Richard. 1995. "Learning to Be Local in Belize: Global Systems of Cultural Difference." In *Worlds Apart: Modernity through the Prism of the Local*, edited by Daniel Miller, 110–133. London: Routledge.

Wilk, Richard. 1999. "'Real Belizean Food': Building Local Identity in the Transnational Caribbean." *American Anthropologist* 101 (2): 244–255.

Wilk, Richard. 2005. "Colonialism and Wildlife in Belize." *Belizean Studies* 27 (2): 4–12.

Wilk, Richard. 2006. *Home Cooking in the Global Village: Caribbean Food from Buccaneers to Ecotourists*. London: Routledge.

Wilk, Richard, and Livia Barbosa. 2012. *Rice and Beans: A Unique Dish in a Hundred Places*. London: Bloomsbury.

Wilk, Richard, and Stephen Miller. 1997. "Some Methodological Issues in Counting Communities and Households." *Human Organization* 56 (1): 64–70.

Williams, Brackette F. 1991. *Stains on My Name, War in My Veins: Guyana and the Politics of Cultural Struggle*. Durham, N.C.: Duke University Press.

Williams, Patricia. 2013. "The Monsterization of Trayvon Martin." *The Nation*, August 19.

Williams, Raymond. 1973. *The Country and the City*. New York: Oxford University Press.

Wilson, Arthur. 1936. "The Logwood Trade in the Seventeenth and Eighteenth Centuries." In *Essays in the History of Modern Europe*, edited by Donald McKay, 1–15. New York: Harper and Row.

Wilson, Elvinet S. 2009. "What It Means to Become a United States American: Afro-Caribbean Immigrants' Constructions of American Citizenship and Experience of Cultural Transition." *Journal of Ethnographic and Qualitative Research* 3 (3): 196–204.

Wilson, Peter. 1973. *Crab Antics: The Social Anthropology of English-Speaking Negro Societies of the Caribbean*. New Haven, Conn.: Yale University Press.

WinklerPrins, Antoinette, and Perpetuo de Sousa. 2009. "House Lot Gardens as Living Space in the Brazilian Amazon." *Focus on Geography* 52 (3/4): 31–38.

Woods, Louis A., Joseph M. Perry, and Jeffrey W. Steagall.1997. "The Composition and Distribution of Ethnic Groups in Belize: Immigration and Emigration Patterns, 1980–1991." *Latin American Research Review* 32 (3): 63–88.

World Travel and Tourism Council. 2017. Belize Tourism and Travel: Economic Impact 2017. Accessed February 17, 2017. https://www.wttc.org/-/media/files/reports/economic-impact-research/countries-2017/belize2017.pdf.

Wright, A. S. 1959. *Land in British Honduras: A Report of the British Honduras Land Use Survey Team.* London: Her Majesty's Stationery Office.

Wynter, Sylvia. 1971. "Novel and History, Plot and Plantation." *Savacou* 5: 95–102.

Wynter, Sylvia. 2000. "The Re-Enchantment of Humanism: An Interview with David Scott." *Small Axe* 8: 119–207.

Wynter, Sylvia. 2003, "Unsettling the Coloniality of Being/Power/Truth/Freedom: Towards the Human, After Man, Its Overrepresentation—An Argument." *CR: The New Centennial Review* 3 (3): 257–337.

Wynter, Sylvia, and Katherine McKittrick, 2015. "Chapter 2: Unparalleled Catastrophe for our Species? Or to Give Humanness a Different Future: Conversations." In *Sylvia Wynter: On Being Human as Praxis*, edited by Katherine McKittrick, 9–89. Durham, N.C.: Duke University Press.

Zarger, Rebecca K., and John R. Stepp. 2004. "Reports: Persistence of Botanical Knowledge among Tzeltal Maya Children." *Current Anthropology* 45 (3): 413–418.

Index

Figures and notes are indicated by *"f" and "n"* following the page numbers.

About the Author

MELISSA A. JOHNSON is a professor of anthropology, environmental studies and race and ethnicity studies at Southwestern University. She has conducted ethnographic research in Belize and within the Belizean diaspora since 1990, and is married into the community and has raised Belizean American sons. She has published articles on her research in Belize in *Environmental History* and *Anthropological Quarterly*, among other journals. Her research and teaching interests focus on making sense of human-environmental relations and processes of racialization.